"After utilizing toolkits from The Art of Service, I was able to identify threats within my organization to which I was completely unaware. Using my team's knowledge as a competitive advantage, we now have superior systems that save time and energy."

"As a new Chief Technology Officer, I was feeling unprepared and inadequate to be successful in my role. I ordered an IT toolkit Sunday night and was prepared Monday morning to shed light on areas of improvement within my organization. I no longer felt overwhelmed and intimidated, I was excited to share what I had learned."

"I used the questionnaires to interview members of my team. I never knew how many insights we could produce collectively with our internal knowledge."

"I usually work until at least 8pm on weeknights. The Art of Service questionnaire saved me so much time and worry that Thursday night I attended my son's soccer game without sacrificing my professional obligations."

"After purchasing The Art of Service toolkit, I was able to identify areas where my company was not in compliance that could have put my job at risk. I looked like a hero when I proactively educated my team on the risks and presented a solid solution."

"I spent months shopping for an external consultant before realizing that The Art of Service would allow my team to consult themselves! Not only did we save time not catching a consultant up to speed, we were able to keep our company information and industry secrets confidential."

"Everyday there are new regulations and processes in my industry. The Art of Service toolkit has kept me ahead by using AI technology to constantly update the toolkits and address emerging needs."

"I customized The Art of Service toolkit to focus specifically on the concerns of my role and industry. I didn't have to waste time with a generic self-help book that wasn't tailored to my exact situation."

"Many of our competitors have asked us about our secret sauce. When I tell them it's the knowledge we have in-house, they never believe me. Little do they know The Art of Service toolkits are working behind the scenes."

"One of my friends hired a consultant who used the knowledge gained working with his company to advise their competitor. Talk about a competitive disadvantage! The Art of Service allowed us to keep our knowledge from walking out the door along with a huge portion of our budget in consulting fees."

"Honestly, I didn't know what I didn't know. Before purchasing The Art of Service, I didn't realize how many areas of my business needed to be refreshed and improved. I am so relieved The Art of Service was there to highlight our blind spots."

"Before The Art of Service, I waited eagerly for consulting company reports to come out each month. These reports kept us up to speed but provided little value because they put our competitors on the same playing field. With The Art of Service, we have uncovered unique insights to drive our business forward."

"Instead of investing extensive resources into an external consultant, we can spend more of our budget towards pursuing our company goals and objectives…while also spending a little more on corporate holiday parties."

"The risk of our competitors getting ahead has been mitigated because The Art of Service has provided us with a 360-degree view of threats within our organization before they even arise."

HR Analytics
Complete Self-Assessment Guide

https://theartofservice.com
support@theartofservice.com

Table of Contents

About The Art of Service

The Art of Service, Business Process Architects since 2000, is dedicated to helping stakeholders achieve excellence.

Defining, designing, creating, and implementing a process to solve a stakeholders challenge or meet an objective is the most valuable role… In EVERY group, company, organization and department.

Unless you're talking a one-time, single-use project, there should be a process. Whether that process is managed and implemented by humans, AI, or a combination of the two, it needs to be designed by someone with a complex enough perspective to ask the right questions.

Someone capable of asking the right questions and step back and say, 'What are we really trying to accomplish here? And is there a different way to look at it?'

With The Art of Service's Self-Assessments, we empower people who can do just that — whether their title is marketer, entrepreneur, manager, salesperson, consultant, Business Process Manager, executive assistant, IT Manager, CIO etc... —they are the people who rule the future. They are people who watch the process as it happens, and ask the right questions to make the process work better.

Contact us when you need any support with this Self-Assessment and any help with templates, blue-prints and examples of standard documents you might need:

https://theartofservice.com
support@theartofservice.com

Included Resources - how to access

Included with your purchase of the book is the HR Analytics

Self-Assessment Spreadsheet Dashboard which contains all questions and Self-Assessment areas and auto-generates insights, graphs, and project RACI planning - all with examples to get you started right away.

How? Simply send an email to
access@theartofservice.com
with this books' title in the subject to get the HR Analytics Self Assessment Tool right away.

INCLUDES LIFETIME SELF ASSESSMENT UPDATES

Every self assessment comes with Lifetime Updates and Lifetime Free Updated Books. Lifetime Updates is an industry-first feature which allows you to receive verified self assessment updates, ensuring you always have the most accurate information at your fingertips.

Get it now- you will be glad you did - do it now, before you forget.

Send an email to **access@theartofservice.com** with this books' title in the subject to get the HR Analytics Self Assessment Tool right away.

Purpose of this Self-Assessment

This Self-Assessment has been developed to improve understanding of the requirements and elements of HR Analytics, based on best practices and standards in business process architecture, design and quality management.

It is designed to allow for a rapid Self-Assessment to determine how closely existing management practices and procedures correspond to the elements of the Self-Assessment.

The criteria of requirements and elements of HR Analytics have been rephrased in the format of a Self-Assessment questionnaire, with a seven-criterion scoring system, as explained in this document.

In this format, even with limited background knowledge of HR Analytics, a manager can quickly review existing operations to determine how they measure up to the standards. This in turn can serve as the starting point of a 'gap analysis' to identify management tools or system elements that might usefully be implemented in the organization to help improve overall performance.

How to use the Self-Assessment

On the following pages are a series of questions to identify to what extent your HR Analytics initiative is complete in comparison to the requirements set in standards.

To facilitate answering the questions, there is a space in front of each question to enter a score on a scale of '1' to '5'.

1 Strongly Disagree

2 Disagree

3 Neutral

4 Agree

5 Strongly Agree

Read the question and rate it with the following in front of mind:

'In my belief,
the answer to this question is clearly defined'.

There are two ways in which you can choose to interpret this statement;

1. how aware are you that the answer to the question is clearly defined
2. for more in-depth analysis you can choose to gather evidence and confirm the answer to the question. This obviously will take more time, most Self-Assessment users opt for the first way to interpret the question and dig deeper later on based on the outcome of the overall Self-Assessment.

A score of '1' would mean that the answer is not clear at all, where a '5' would mean the answer is crystal clear and defined. Leave emtpy when the question is not applicable or you don't want to answer it, you can skip it without affecting your score. Write your score in the space provided.

After you have responded to all the appropriate statements in each section, compute your average score for that section, using the formula provided, and round to the nearest tenth. Then transfer to the corresponding spoke in the HR Analytics Scorecard on the second next page of the Self-Assessment.

Your completed HR Analytics Scorecard will give you a clear presentation of which HR Analytics areas need attention.

HR Analytics Scorecard Example

Example of how the finalized Scorecard can look like:

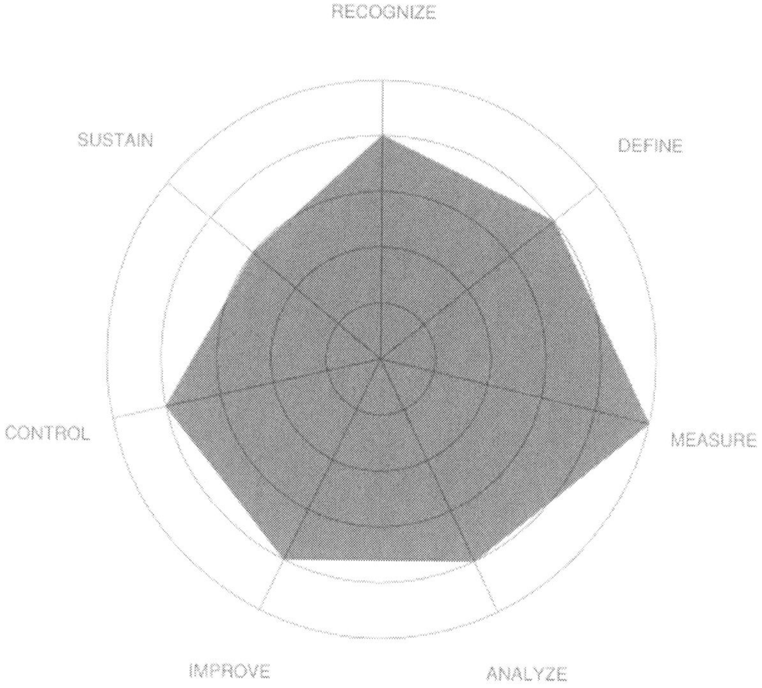

HR Analytics Scorecard

Your Scores:

BEGINNING OF THE SELF-ASSESSMENT:

CRITERION #1: RECOGNIZE

INTENT: Be aware of the need for change. Recognize that there is an unfavorable variation, problem or symptom.

In my belief, the answer to this question is clearly defined:

5 Strongly Agree

4 Agree

3 Neutral

2 Disagree

1 Strongly Disagree

1. Who else hopes to benefit from it?
<--- Score

2. Are the roles and responsibilities clearly identified and are responsibilities executed effectively?
<--- Score

3. Where do you shine and what needs work?

<--- Score

4. Does your team culture need a makeover?
<--- Score

5. What do needs and goals have to do with yours?
<--- Score

6. What hrm issues are critical to your organizations long term sustainability?
<--- Score

7. What are the Workforce issues/challenges at hand?
<--- Score

8. Will some divisions need to be reorganized to meet business needs and strategic objectives?
<--- Score

9. Why do other organizations need governance?
<--- Score

10. Will your seniority be recognized by the successor organization?
<--- Score

11. What does HR Analytics success mean to the stakeholders?
<--- Score

12. What core capabilities do you need to win in your markets?
<--- Score

13. Why is leadership a perennial issue?

<--- Score

14. Which issues related to employees involvement in HRM will play an important role in the future?
<--- Score

15. How do you identify your top performers?
<--- Score

16. Does it offer the functionality you need to succeed?
<--- Score

17. How do you know if your teams culture is in need of an overhaul a makeover?
<--- Score

18. Does the problem lie with the employee?
<--- Score

19. How does an HRM strategy reinforce and deal with the need for change and innovation?
<--- Score

20. What hrm functions do you need for sustainability?
<--- Score

21. Do you need to expand your workforce levels to meet business demand?
<--- Score

22. What are your current pressing business issues?
<--- Score

23. What kinds of issues could make most of the customers angry?
<--- Score

24. What problems are you facing and how do you consider HR Analytics will circumvent those obstacles?
<--- Score

25. What are the stakeholder objectives to be achieved with HR Analytics?
<--- Score

26. Does the system issue referrals?
<--- Score

27. How are you going to measure success?
<--- Score

28. How are you seeing total rewards evolve to meet the changing needs of the workforce?
<--- Score

29. Does the client need reduced downtime through traditional recovery methods?
<--- Score

30. What resources will you need for maintenance?
<--- Score

31. Have the specialized skills or knowledge needed for the job?
<--- Score

32. Does hris identify unfilled positions accurately?
<--- Score

33. What skills will you need in the future?
<--- Score

34. How many employees are needed in each job area?
<--- Score

35. How do you identify critical talent segments?
<--- Score

36. How much are sponsors, customers, partners, stakeholders involved in HR Analytics? In other words, what are the risks, if HR Analytics does not deliver successfully?
<--- Score

37. Is hrm unconsciously fostering an environment in which employees no longer identify with organizational goals, values and activities?
<--- Score

38. What kind of people do you need?
<--- Score

39. What skills you need now and in the future?
<--- Score

40. What are the key people issues driving or hindering strategic agenda?
<--- Score

41. Are there any specific expectations or concerns about the HR Analytics team, HR Analytics itself?
<--- Score

42. What new kinds of talent will be needed?

<--- Score

43. What resources will you need now?
<--- Score

44. What recognition and rewards are needed to attract, motivate, and retain the employees you need?
<--- Score

45. What situation(s) led to this HR Analytics Self Assessment?
<--- Score

46. What story does your organization need?
<--- Score

47. As a sponsor, customer or management, how important is it to meet goals, objectives?
<--- Score

48. How do you know when a culture shift is needed?
<--- Score

49. What percentage of your key positions have identified successors?
<--- Score

50. Are key stakeholders and organization leadership apprised of the concrete need for change and action?
<--- Score

51. Why should you consider the need to attract talent when there are so many people looking for

work?
<--- Score

52. What are the expected benefits of HR Analytics to the stakeholder?
<--- Score

53. What would happen if HR Analytics weren't done?
<--- Score

54. How are the HR Analytics's objectives aligned to the group's overall stakeholder strategy?
<--- Score

55. Have you identified the right tools and technology?
<--- Score

56. Do you readily determine the training needs of your employees?
<--- Score

Add up total points for this section:
_____ = Total points for this section

Divided by: ______ (number of statements answered) = ______ Average score for this section

Transfer your score to the HR Analytics Index at the beginning of the Self-Assessment.

CRITERION #2: DEFINE:

INTENT: Formulate the stakeholder problem. Define the problem, needs and objectives.

In my belief, the answer to this question is clearly defined:

5 Strongly Agree

4 Agree

3 Neutral

2 Disagree

1 Strongly Disagree

1. Has a project plan, Gantt chart, or similar been developed/completed?
<--- Score

2. Have you defined the future of reporting for your organization?
<--- Score

3. Has the HR Analytics work been fairly and/or

equitably divided and delegated among team members who are qualified and capable to perform the work? Has everyone contributed?
<--- Score

4. Are different versions of process maps needed to account for the different types of inputs?
<--- Score

5. How is the team tracking and documenting its work?
<--- Score

6. Has the direction changed at all during the course of HR Analytics? If so, when did it change and why?
<--- Score

7. When is/was the HR Analytics start date?
<--- Score

8. How many cases where the model predicts retention and the employee stays?
<--- Score

9. Has anyone else (internal or external to the group) attempted to solve this problem or a similar one before? If so, what knowledge can be leveraged from these previous efforts?
<--- Score

10. Will team members regularly document their HR Analytics work?
<--- Score

11. What are the compelling stakeholder reasons for embarking on HR Analytics?

<--- Score

12. Does the position require specialized or unique skills or knowledge that may be difficult to replace?
<--- Score

13. Is the current 'as is' process being followed? If not, what are the discrepancies?
<--- Score

14. Are stakeholder processes mapped?
<--- Score

15. Are customer(s) identified and segmented according to their different needs and requirements?
<--- Score

16. When is the estimated completion date?
<--- Score

17. What critical content must be communicated – who, what, when, where, and how?
<--- Score

18. Is your highly experienced and skilled talent deployed in positions that require skills?
<--- Score

19. What are the dynamics of the communication plan?
<--- Score

20. What percentage of the workforce has a defined career path?
<--- Score

21. If substitutes have been appointed, have they been briefed on the HR Analytics goals and received regular communications as to the progress to date?
<--- Score

22. How many hours per week define full-time?
<--- Score

23. Is full participation by members in regularly held team meetings guaranteed?
<--- Score

24. What would be the goal or target for a HR Analytics's improvement team?
<--- Score

25. Has/have the customer(s) been identified?
<--- Score

26. Can required skills be obtained through sharing resources?
<--- Score

27. What baseline technology will be required to support the new service delivery model?
<--- Score

28. Has a team charter been developed and communicated?
<--- Score

29. How many cases does the model predict attrition and the employee quits?
<--- Score

30. Is a fully trained team formed, supported, and committed to work on the HR Analytics improvements?
<--- Score

31. Are customers identified and high impact areas defined?
<--- Score

32. Is there regularly 100% attendance at the team meetings? If not, have appointed substitutes attended to preserve cross-functionality and full representation?
<--- Score

33. Has the improvement team collected the 'voice of the customer' (obtained feedback – qualitative and quantitative)?
<--- Score

34. What customer feedback methods were used to solicit their input?
<--- Score

35. How was the 'as is' process map developed, reviewed, verified and validated?
<--- Score

36. Have the customer needs been translated into specific, measurable requirements? How?
<--- Score

37. Is HR Analytics linked to key stakeholder goals and objectives?
<--- Score

38. Is there a critical path to deliver HR Analytics results?
<--- Score

39. Is data collected and displayed to better understand customer(s) critical needs and requirements.
<--- Score

40. Will team members perform HR Analytics work when assigned and in a timely fashion?
<--- Score

41. Is the improvement team aware of the different versions of a process: what they think it is vs. what it actually is vs. what it should be vs. what it could be?
<--- Score

42. How do you check the business has access to all the information it requires at the right time?
<--- Score

43. How often are the team meetings?
<--- Score

44. Is HR Analytics currently on schedule according to the plan?
<--- Score

45. What key stakeholder process output measure(s) does HR Analytics leverage and how?
<--- Score

46. What are the boundaries of the scope? What is in bounds and what is not? What is the start point? What is the stop point?

<--- Score

47. Is there a completed SIPOC representation, describing the Suppliers, Inputs, Process, Outputs, and Customers?
<--- Score

48. How will variation in the actual durations of each activity be dealt with to ensure that the expected HR Analytics results are met?
<--- Score

49. Are there different segments of customers?
<--- Score

50. Has a high-level 'as is' process map been completed, verified and validated?
<--- Score

51. Who are the HR Analytics improvement team members, including Management Leads and Coaches?
<--- Score

52. How does the HR Analytics manager ensure against scope creep?
<--- Score

53. Does the team have regular meetings?
<--- Score

54. Do current hris meet the requirements of HRM?
<--- Score

55. When are meeting minutes sent out? Who is on the distribution list?

<--- Score

56. Does duty require access to and/or confidentiality of information?
<--- Score

57. How will the HR Analytics team and the group measure complete success of HR Analytics?
<--- Score

58. Do the problem and goal statements meet the SMART criteria (specific, measurable, attainable, relevant, and time-bound)?
<--- Score

59. What types of skills and positions will be required or no longer required?
<--- Score

60. Is there a completed, verified, and validated high-level 'as is' (not 'should be' or 'could be') stakeholder process map?
<--- Score

61. How did the HR Analytics manager receive input to the development of a HR Analytics improvement plan and the estimated completion dates/times of each activity?
<--- Score

62. How do you keep key subject matter experts in the loop?
<--- Score

63. Are there any constraints known that bear on the ability to perform HR Analytics work? How is the team

addressing them?
<--- Score

64. How do you define innovation within human resources?
<--- Score

65. What are the Roles and Responsibilities for each team member and its leadership? Where is this documented?
<--- Score

66. Is the HR Analytics scope manageable?
<--- Score

67. Is the team adequately staffed with the desired cross-functionality? If not, what additional resources are available to the team?
<--- Score

68. Is the team sponsored by a champion or stakeholder leader?
<--- Score

69. Is there a HR Analytics management charter, including stakeholder case, problem and goal statements, scope, milestones, roles and responsibilities, communication plan?
<--- Score

70. Are team charters developed?
<--- Score

71. What skills and competencies are required to fill the critical jobs?
<--- Score

72. Is the team equipped with available and reliable resources?
<--- Score

73. What specifically is the problem? Where does it occur? When does it occur? What is its extent?
<--- Score

74. Has everyone on the team, including the team leaders, been properly trained?
<--- Score

75. Is the team formed and are team leaders (Coaches and Management Leads) assigned?
<--- Score

76. What constraints exist that might impact the team?
<--- Score

77. What are the rough order estimates on cost savings/opportunities that HR Analytics brings?
<--- Score

78. Are improvement team members fully trained on HR Analytics?
<--- Score

Add up total points for this section:
_____ = Total points for this section

Divided by: ______ (number of statements answered) = ______
Average score for this section

Transfer your score to the HR Analytics Index at the beginning of the Self-Assessment.

CRITERION #3: MEASURE:

INTENT: Gather the correct data. Measure the current performance and evolution of the situation.

In my belief, the answer to this question is clearly defined:

5 Strongly Agree

4 Agree

3 Neutral

2 Disagree

1 Strongly Disagree

1. What is the capacity to implement and maintain the measures?
<--- Score

2. What are the factors driving an analytics approach?
<--- Score

3. What kinds of human capital analytics are

utilised and where?
<--- Score

4. Does your organization recognize the value of analytically driven insight?
<--- Score

5. How much data do you currently use for analysis?
<--- Score

6. Do your organizations processes explicitly accommodate analytics?
<--- Score

7. How have leadership development programs impacted productivity, profitability, and employee satisfaction?
<--- Score

8. How and why does workforce equity have an impact?
<--- Score

9. Can a method be used to analyze many different jobs?
<--- Score

10. What is the difference between metrics and analytics?
<--- Score

11. How are people analytics functions structured?
<--- Score

12. Is there a person or team that is dedicated to

people analytics?
<--- Score

13. Should employees be the focus of remedial action?
<--- Score

14. Which method should job analysts use?
<--- Score

15. What is your organizational impact of the shift to cloud based HRIS?
<--- Score

16. Are you feeling the business impact of downtime caused by unresolved service issues?
<--- Score

17. Where does your organization best apply analytics to improve talent management?
<--- Score

18. What are human resources analytics?
<--- Score

19. What do you measure about your workforce?
<--- Score

20. What are the most important areas to expand your workforce analytics?
<--- Score

21. How are other organizations progressing in the field of human capital analytics?
<--- Score

22. What are the current analytics practices in talent management?
<--- Score

23. What kind of work is being done in talent analytics?
<--- Score

24. What was the impact on the business?
<--- Score

25. What is impacting that efficiency and what could improve it?
<--- Score

26. How well the current data sources cover the needs of people analytics?
<--- Score

27. Who should deliver your analytics capability?
<--- Score

28. Do you have a standardized approach to workforce analytics projects?
<--- Score

29. Are the variance analysis processes meaningful and lead to appropriate corrective actions?
<--- Score

30. What proportion of workers have had a formal job analysis?
<--- Score

31. Do you show return on investment for your analytics projects?

<--- Score

32. How confident are you in your teams understanding and ability to interpret analytics?
<--- Score

33. Are senior leaders advocates of analytics or sceptics?
<--- Score

34. What is the cost associated with the strategies to address competency or skill gaps?
<--- Score

35. What are the projected future labor costs?
<--- Score

36. How does your organization adopt people analytics?
<--- Score

37. Where is ai already making an impact?
<--- Score

38. Which model do you have to use, the regression analysis or the database?
<--- Score

39. What technologies and tools do you need to support your analytic capability?
<--- Score

40. What could you have done differently to better impact your sphere of influence?
<--- Score

41. What area of analysis would most likely hold the required information?
<--- Score

42. What tools are used for analysis /reporting?
<--- Score

43. Are you using the right human resource measures?
<--- Score

44. What type of analytics do you need?
<--- Score

45. What is business analytics and which tools are used for analysis?
<--- Score

46. What type of employee analysis do you currently conduct?
<--- Score

47. How might impact the actions of your organization?
<--- Score

48. What is the appetite on the ground for talent analytics and big data?
<--- Score

49. How can people analytics make HRM more evidence based?
<--- Score

50. What is talent analytics and why do you measure?

<--- Score

51. How can the information acquired by a job analysis be used in the development of HRM programs?
<--- Score

52. What kind of analytics are you using for pay equity analysis?
<--- Score

53. Do you have a strategy for prioritizing workforce analytics?
<--- Score

54. What core metrics/analysis should you use to inform data consumers?
<--- Score

55. How much do legal problems cost?
<--- Score

56. Are you focusing on value creation in every job?
<--- Score

57. Do you know which measures can make your talent management strategy more effective?
<--- Score

58. How do analytics, metrics and benchmarking relate to one another?
<--- Score

59. Where should you focus in reskilling and upskilling your workforce?

<--- Score

60. Have you quantified the level of risk or size of missed opportunity?
<--- Score

61. What is required to set up and run an effective talent analytics function?
<--- Score

62. How will job redesign impact the current classification and compensation system?
<--- Score

63. What is the most sophisticated type of analytics your business unit regularly employs?
<--- Score

64. How will you quantify benefits gained from your investment in analytics?
<--- Score

65. What are the main development items for progressing in analytics?
<--- Score

66. Are you having problems controlling the total cost of ownership across your end user environment?
<--- Score

67. How do you measure contingent workers performance?
<--- Score

68. Does your organization invest in good quality

data and analysis?
<--- Score

69. What are the design aspects of jobs that make the biggest impact on business outcomes?
<--- Score

70. What are the byproducts of job analysis?
<--- Score

71. Are any line item costs increasing at a disproportionate rate to revenue?
<--- Score

72. How do you transition to more advanced analytics capabilities?
<--- Score

73. How and when will you measure success?
<--- Score

74. What are the key steps you can take to build your people analytics strategy?
<--- Score

75. What is business intelligence and analytics?
<--- Score

76. How is the value of the IT investment/return measured?
<--- Score

77. What is driving high employee turnover, and how much is it costing your business?
<--- Score

78. Is your organizations data structured and available in a way that supports people analytics?
<--- Score

79. How is workforce analytics aligned to the business objectives of your organization?
<--- Score

80. How long has your organization had a person or team dedicated to people analytics?
<--- Score

81. Which roles or skill sets have the highest strategic impact on the goals of your organization?
<--- Score

82. What tools do you use for analytics?
<--- Score

83. Do you have the right analytics technology for your needs?
<--- Score

84. What is the impact of manager skills on employee relation issues or team performance?
<--- Score

85. What technologies are in place to support human capital analytics initiatives?
<--- Score

86. Does your organization have the capabilities to meet all its anticipated analytics needs?
<--- Score

87. What kind of organization and specific skills

are needed to support an analytics capability?
<--- Score

88. What is the impact of globalisation on the International HRM?
<--- Score

89. What does a people analytics strategy look like?
<--- Score

90. Where are you on the analytics journey?
<--- Score

91. Do you have a skilled workforce analytics leader?
<--- Score

92. What is driving the rise of analytics?
<--- Score

93. How do people impact success in your part of the business?
<--- Score

94. Is it cost effective and/or good business practice to contract out for special competencies?
<--- Score

95. What is the cost impact across recruiting, training, and lost productivity for each early departure?
<--- Score

96. How much is absenteeism costing your organization?

<--- Score

97. What impact does human capital have on financial performance?
<--- Score

98. Does the system capture information on time, or units of measurement worked, to determine gross pay?
<--- Score

99. What picture best describes analytics?
<--- Score

100. Do you have stakeholders who want to solve problems analytically?
<--- Score

101. Do you model your workforce to optimize cost, profit, and productivity?
<--- Score

102. How do you advance from reporting of transaction data to advanced analytics?
<--- Score

103. How do you hire, retain, communicate and organize your analytic talent?
<--- Score

104. What are the costs versus the benefits of the recruitment strategy?
<--- Score

105. How do you link the business impact of your project to your organizations success?

<--- Score

106. How many ftes are on the people analytics team?
<--- Score

107. What analytical procedures are used in the analysis of available data?
<--- Score

108. How well the current data sources cover the needs of human capital analytics?
<--- Score

109. Which trainer or training program has had the biggest impact in terms of improving productivity?
<--- Score

110. Who owns analytics in your organization?
<--- Score

111. Do you have a vision and mission for workforce analytics?
<--- Score

112. What is your average cost per hire?
<--- Score

113. How critical are workforce analytics to the success of your business?
<--- Score

114. What are the likely implementation costs for a new system?
<--- Score

115. What will workforce analytics do for your firm?
<--- Score

116. Is the business ready for analytics?
<--- Score

117. Do you have senior sponsors who will champion analytics?
<--- Score

118. How do you build an effective analytics centre of excellence?
<--- Score

119. What is the leading way to improve the impact of decision support tools?
<--- Score

120. Where should data analytics sit?
<--- Score

121. How do you develop an approach to talent analytics?
<--- Score

122. Did the analysis include an intention to treat analysis?
<--- Score

123. Which create the greatest need for analytical skills in your organization?
<--- Score

124. What are the predictive factors that cause top

performers to deliver better results?
<--- Score

125. Where is your organization on the measurement continuum?
<--- Score

126. What are the startup costs, resources, and supervision?
<--- Score

127. What is the true cost impact of turnover?
<--- Score

128. Do you understand what the term workforce analytics means?
<--- Score

129. What should your organization look like to support the reporting and analytics needs of stakeholders?
<--- Score

130. Which functions have strong analytics capabilities?
<--- Score

Add up total points for this section:
_____ = Total points for this section

Divided by: ______ (number of statements answered) = ______ Average score for this section

Transfer your score to the HR Analytics Index at the beginning of the Self-

Assessment.

CRITERION #4: ANALYZE:

INTENT: Analyze causes, assumptions and hypotheses.

In my belief, the answer to this question is clearly defined:

5 Strongly Agree

4 Agree

3 Neutral

2 Disagree

1 Strongly Disagree

1. Is the HR Analytics process severely broken such that a re-design is necessary?
<--- Score

2. Are gaps between current performance and the goal performance identified?
<--- Score

3. How do you ensure security of personal data?
<--- Score

4. What were the crucial 'moments of truth' on the process map?
<--- Score

5. How do you migrate data successfully?
<--- Score

6. Who is responsible for assuring proper use of the data?
<--- Score

7. How do you gain new customer insight from social media data?
<--- Score

8. Is an increase in absence driven by increasing stress levels?
<--- Score

9. What tools were used to narrow the list of possible causes?
<--- Score

10. How reliable and valid is the data?
<--- Score

11. Will users have access to all the data on the system or will the users access be restricted?
<--- Score

12. How do you ensure accuracy and completeness of data?
<--- Score

13. Were there any improvement opportunities

identified from the process analysis?
<--- Score

14. Do the new initiatives, systems, processes and practices require new skill sets?
<--- Score

15. Does corporate culture influence access and use of data?
<--- Score

16. Do other systems share data or have access to the data in the system?
<--- Score

17. How should workforce equity drive change?
<--- Score

18. Was a cause-and-effect diagram used to explore the different types of causes (or sources of variation)?
<--- Score

19. What experiences drive employee engagement?
<--- Score

20. Did you have sufficient development and growth opportunities?
<--- Score

21. Is big data a threat or an opportunity?
<--- Score

22. What quality tools were used to get through the analyze phase?
<--- Score

23. What are the retention periods of the data in the system?
<--- Score

24. Are the skills of employees being developed quickly enough to become effective?
<--- Score

25. What are some best practices to ensure data integrity?
<--- Score

26. Are key positions filled quickly after a vacancy?
<--- Score

27. Have you identified the challenges and gaps with your current processes?
<--- Score

28. Have any additional benefits been identified that will result from closing all or most of the gaps?
<--- Score

29. What is good universal talent data?
<--- Score

30. What data quality issues are there?
<--- Score

31. What were the financial benefits resulting from any 'ground fruit or low-hanging fruit' (quick fixes)?
<--- Score

32. Is it ethical to collect a wide range of data on the employees?

<--- Score

33. What does the data say about the performance of the stakeholder process?
<--- Score

34. What did the team gain from developing a sub-process map?
<--- Score

35. What is the cost of poor quality as supported by the team's analysis?
<--- Score

36. Are procedures in place to promptly process other forms of ordered relief?
<--- Score

37. Do you export data from your system?
<--- Score

38. What kind of leaders can drive high engagement?
<--- Score

39. What data do managers require to enable substantial improvements in the workforce?
<--- Score

40. What is appetite and aptitude for data?
<--- Score

41. Which hrm activities/processes can be digitized and why?
<--- Score

42. Is the performance gap determined?
<--- Score

43. Do you have sufficient data coverage?
<--- Score

44. What data sources are integrated into a single bi tool?
<--- Score

45. How coherent is the human capital data governance?
<--- Score

46. How do staff find opportunities to develop new knowledge and skills for currnet jobs?
<--- Score

47. Is there a formal process for handling emergency spending and integrating results into existing capital budgets?
<--- Score

48. Can changes be made to all captured data?
<--- Score

49. Does the system process leave for each reported leave type at the end of each effective pay period?
<--- Score

50. What are the data quality risks?
<--- Score

51. Were any designed experiments used to generate additional insight into the data analysis?

<--- Score

52. Do you have access to workforce/people data that is produced by your organization?
<--- Score

53. Do you easily search for all data relating to a particular individual?
<--- Score

54. Who has access to people data for decision making purposes?
<--- Score

55. Did the recruiting department assist you to your satisfaction throughout the hiring process?
<--- Score

56. Do you lose talent during the recruitment process?
<--- Score

57. Does it have the proper data governance practices in place?
<--- Score

58. Is there a process for employees to lodge complaints?
<--- Score

59. What to do with all of the data?
<--- Score

60. What are the opportunities that you see for your organizations workforce?
<--- Score

61. How will the new data be verified for relevance and accuracy?
<--- Score

62. How are the recruitment, screening, and selection processes documented?
<--- Score

63. Is data and process analysis, root cause analysis and quantifying the gap/opportunity in place?
<--- Score

64. What conclusions were drawn from the team's data collection and analysis? How did the team reach these conclusions?
<--- Score

65. How do you successfully govern a vast amount of data?
<--- Score

66. How are managers using people data in roles?
<--- Score

67. Have the problem and goal statements been updated to reflect the additional knowledge gained from the analyze phase?
<--- Score

68. How do new hires view the onboarding process?
<--- Score

69. Will the new data be placed in the individuals record?

<--- Score

70. What kind of connections can be seen between concepts in the overall data?
<--- Score

71. Is your process centrally managed or distributed?
<--- Score

72. What tools were used to generate the list of possible causes?
<--- Score

73. Which programs drive better workforce performance?
<--- Score

74. Were Pareto charts (or similar) used to portray the 'heavy hitters' (or key sources of variation)?
<--- Score

75. Is data considered an asset by leadership?
<--- Score

76. Is the gap/opportunity displayed and communicated in financial terms?
<--- Score

77. Which groups should receive data and insights?
<--- Score

78. What is preventing organizations from leveraging data effectively?
<--- Score

79. What will you do with all of the data?
<--- Score

80. Is there a separate reporting environment or data mart?
<--- Score

81. What was the purpose of the collected data?
<--- Score

82. Was a detailed process map created to amplify critical steps of the 'as is' stakeholder process?
<--- Score

83. How do you get the most value out of internal and external data?
<--- Score

84. Who is getting access to the data and in what form?
<--- Score

85. Who will have access to the data in the system?
<--- Score

86. What tools and systems are necessary for data driven decisions?
<--- Score

87. Did any value-added analysis or 'lean thinking' take place to identify some of the gaps shown on the 'as is' process map?
<--- Score

88. What insights will drive value for your

organization?
<--- Score

89. What are the data elements needed?
<--- Score

90. Did any additional data need to be collected?
<--- Score

91. What services are offered to ensure a smooth implementation process?
<--- Score

92. What does the optimization process look like?
<--- Score

93. Will the system derive new data?
<--- Score

94. Why is data & measurement so important to the business?
<--- Score

95. How is access to the data by a user determined?
<--- Score

96. Are there adequate geographic data to assist in design projects accurately?
<--- Score

97. Do you have a high exposure to virus, theft, lose of devices and data?
<--- Score

98. How was the detailed process map generated,

verified, and validated?
<--- Score

99. How do you ensure good quality data?
<--- Score

100. Are you fair in your selection process?
<--- Score

101. What employee data and which KPIs are most valuable to your organization to measure?
<--- Score

102. What are the revised rough estimates of the financial savings/opportunity for HR Analytics improvements?
<--- Score

Add up total points for this section:
_____ = Total points for this section

Divided by: ______ (number of statements answered) = ______ Average score for this section

Transfer your score to the HR Analytics Index at the beginning of the Self-Assessment.

CRITERION #5: IMPROVE:

INTENT: Develop a practical solution. Innovate, establish and test the solution and to measure the results.

In my belief, the answer to this question is clearly defined:

5 Strongly Agree

4 Agree

3 Neutral

2 Disagree

1 Strongly Disagree

1. Which practical developments in employee involvement in HRM are likely to occur in the future?
<--- Score

2. What risks/benefits are associated with your approach to pay for performance?
<--- Score

3. Do you have pay compliance risks that need to be addressed?
<--- Score

4. What is the implementation plan?
<--- Score

5. Do the competencies actually map to success?
<--- Score

6. What is your workforce productivity and is it improving?
<--- Score

7. What lessons, if any, from a pilot were incorporated into the design of the full-scale solution?
<--- Score

8. Is there a small-scale pilot for proposed improvement(s)? What conclusions were drawn from the outcomes of a pilot?
<--- Score

9. Do you truly understand how your organization makes money?
<--- Score

10. How does that play out in reality based on your organizations actual financial results?
<--- Score

11. Is the optimal solution selected based on testing and analysis?
<--- Score

12. How will the team or the process owner(s) monitor

the implementation plan to see that it is working as intended?
<--- Score

13. Is it improving as a percentage of revenue?
<--- Score

14. Are staff developing themselves in a way that maps personal effectiveness with the achievement of strategic objectives?
<--- Score

15. Does the system check for effectiveness of the evaluation?
<--- Score

16. Are the systems and procedures used to provide accurate customer billing and receivables and collections well developed and applied reasonably and adequately used?
<--- Score

17. What is the percentage increase in employee performance as a result of every one percent increase in pay?
<--- Score

18. Which lines of business or managers are the best developers of talent as measured by the promotions out of departments?
<--- Score

19. Is there a cost/benefit analysis of optimal solution(s)?
<--- Score

20. How will your organization develop worker competencies?
<--- Score

21. Is a solution implementation plan established, including schedule/work breakdown structure, resources, risk management plan, cost/budget, and control plan?
<--- Score

22. Are the best solutions selected?
<--- Score

23. Is pilot data collected and analyzed?
<--- Score

24. What have you done differently as a result of advice provided by your business partner?
<--- Score

25. What is the biggest risk when making hiring decisions based on statistical behavioral studies?
<--- Score

26. What error proofing will be done to address some of the discrepancies observed in the 'as is' process?
<--- Score

27. What is your approach to paas development?
<--- Score

28. How has your performance management system improved performance in your organization?
<--- Score

29. Does the vendors solution meet the current and future key requirements of your organization?
<--- Score

30. How much do you spend annually per individual on training and development?
<--- Score

31. How is the development of your organization in the near future?
<--- Score

32. Do workers understand the business strategy and how work contributes to organization performance?
<--- Score

33. What tools were used to evaluate the potential solutions?
<--- Score

34. Can movement be implemented through career development programs?
<--- Score

35. Which roles require a long training time to develop the skills for the position?
<--- Score

36. What attendant changes will need to be made to ensure that the solution is successful?
<--- Score

37. Are you getting ROI on your training and development activities?
<--- Score

38. Were any criteria developed to assist the team in testing and evaluating potential solutions?
<--- Score

39. What percentage of your high performers are at high risk for departure?
<--- Score

40. What are the predictive indicators of turnover, and which of your employees are at risk?
<--- Score

41. Describe the design of the pilot and what tests were conducted, if any?
<--- Score

42. Where can talent leaders improve business results?
<--- Score

43. How do you achieve similar business results?
<--- Score

44. Are there any constraints (technical, political, cultural, or otherwise) that would inhibit certain solutions?
<--- Score

45. How much corruption is there in decision making?
<--- Score

46. What does the 'should be' process map/design look like?
<--- Score

47. Are there fundamental changes occurring in your organization as a result of industry trends?
<--- Score

48. Is a contingency plan established?
<--- Score

49. How do you attract, retain and develop skills?
<--- Score

50. What tools were most useful during the improve phase?
<--- Score

51. Do you feel the brief descriptions provided for benefits are easy to understand?
<--- Score

52. Is that leader development program adding value and if so, how much?
<--- Score

53. How much weight in managers variable pay or performance reviews is given to employee development?
<--- Score

54. What decisions are made with it?
<--- Score

55. What is HR Analytics's impact on utilizing the best solution(s)?
<--- Score

56. How does the solution remove the key sources of

issues discovered in the analyze phase?
<--- Score

57. Does your organization have adequate systems for customer billing, accounts receivable, and collections in place to safeguard assets as well as to record, summarize, and report the financial results?
<--- Score

58. Are new and improved process ('should be') maps developed?
<--- Score

59. Is the corporate strategy documented?
<--- Score

60. Are possible solutions generated and tested?
<--- Score

61. What is the risk to the continuity of key positions?
<--- Score

62. Is quality and quantity of work evaluated?
<--- Score

63. What is the team's contingency plan for potential problems occurring in implementation?
<--- Score

64. Which skills/talent groups can be developed internally?
<--- Score

65. Is the implementation plan designed?

<--- Score

66. Has productivity increased as a result of acquiring new skills and knowledge?
<--- Score

67. How much time do you have to identify and develop replacements for employees who are retiring?
<--- Score

68. Where should you open your new Research & Development center or your new office?
<--- Score

69. Does the vendor understand your organization and its challenges?
<--- Score

70. Do you need to evaluate and increase your employer branding investment?
<--- Score

71. Which individual employees across different jobs are in high risk of attrition?
<--- Score

72. How do you attract, hire, grow and retain the right talent to positively contribute to your bottom line through innovative solutions?
<--- Score

73. Are improved process ('should be') maps modified based on pilot data and analysis?
<--- Score

74. Is there a system for performance evaluation?
<--- Score

75. Where are the greatest risks of internal fraud and embezzlement?
<--- Score

76. Which managers cultivate talent and promote employee development?
<--- Score

77. Do you agree with broad hiring policies driving decisions between one individual and another?
<--- Score

78. Who has the authority to develop a new pay program?
<--- Score

79. How will the group know that the solution worked?
<--- Score

80. What are the barriers to optimally achieving the strategy?
<--- Score

81. What tools were used to tap into the creativity and encourage 'outside the box' thinking?
<--- Score

82. What were the underlying assumptions on the cost-benefit analysis?
<--- Score

83. What suggestions do you have for improving

the work experience with your organization?
<--- Score

84. Does a member of management evaluate each termination before it takes place?
<--- Score

85. How to structure decision making for innovation contexts?
<--- Score

86. Was a pilot designed for the proposed solution(s)?
<--- Score

87. How are staff supported in development?
<--- Score

88. How does your brand as an employer develop?
<--- Score

89. How did the team generate the list of possible solutions?
<--- Score

90. What are the key workforce issues and risks you are currently facing?
<--- Score

91. What communications are necessary to support the implementation of the solution?
<--- Score

92. Which departments need improvement?
<--- Score

Add up total points for this section:

_____ = Total points for this section

Divided by: ______ (number of statements answered) = ______ Average score for this section

Transfer your score to the HR Analytics Index at the beginning of the Self-Assessment.

CRITERION #6: CONTROL:

INTENT: Implement the practical solution. Maintain the performance and correct possible complications.

In my belief, the answer to this question is clearly defined:

5 Strongly Agree

4 Agree

3 Neutral

2 Disagree

1 Strongly Disagree

1. What are the critical parameters to watch?
<--- Score

2. Is there documentation that will support the successful operation of the improvement?
<--- Score

3. How many accruing time off plans do you need configured?

<--- Score

4. How will the process owner and team be able to hold the gains?
<--- Score

5. Is there a standardized process?
<--- Score

6. Are you providing data driven learning & development to support non linear career paths?
<--- Score

7. Is knowledge gained on process shared and institutionalized?
<--- Score

8. How many benefit plans will you need?
<--- Score

9. Are the goals of the succession plan still appropriate?
<--- Score

10. Are operating procedures consistent?
<--- Score

11. Is control on corporate level or will line managers and employees get more control?
<--- Score

12. Why does the second legal plan have more employees enrolled in a legal plan?
<--- Score

13. What have you learned about generic

competitive strategy?
<--- Score

14. What is the control/monitoring plan?
<--- Score

15. Who will have the control in the future?
<--- Score

16. What should the next improvement project be that is related to HR Analytics?
<--- Score

17. How will the day-to-day responsibilities for monitoring and continual improvement be transferred from the improvement team to the process owner?
<--- Score

18. Are there documented procedures?
<--- Score

19. How will the process owner verify improvement in present and future sigma levels, process capabilities?
<--- Score

20. What compensation elements are being planned?
<--- Score

21. What is the recommended frequency of auditing?
<--- Score

22. Do you plan to do a talent review annually?
<--- Score

23. Have all major milestones in your organization plan been completed?
<--- Score

24. What does it take to learn the language of business, numbers and financial impact?
<--- Score

25. Are suggested corrective/restorative actions indicated on the response plan for known causes to problems that might surface?
<--- Score

26. How do you rate your organization on areas of human capital analytics in decision making against your industry?
<--- Score

27. Does a troubleshooting guide exist or is it needed?
<--- Score

28. Do certain legal plans have a better mix of benefits and access that employees will need and value?
<--- Score

29. Does it allow monitoring and tracking of workforce performance?
<--- Score

30. Is new knowledge gained imbedded in the response plan?
<--- Score

31. How might the group capture best practices and lessons learned so as to leverage improvements?

<--- Score

32. Is there a recommended audit plan for routine surveillance inspections of HR Analytics's gains?
<--- Score

33. Does your organization have a plan to conduct periodic monitoring to measure progress?
<--- Score

34. How will your organization implement what was learned?
<--- Score

35. What other systems, operations, processes, and infrastructures (hiring practices, staffing, training, incentives/rewards, metrics/dashboards/scorecards, etc.) need updates, additions, changes, or deletions in order to facilitate knowledge transfer and improvements?
<--- Score

36. Do you ask when an employee is planning to retire?
<--- Score

37. How will you evaluate progress in implementing the plan?
<--- Score

38. Does the response plan contain a definite closed loop continual improvement scheme (e.g., plan-do-check-act)?
<--- Score

39. Do you plan for your annual headcount?

<--- Score

40. What quality tools were useful in the control phase?
<--- Score

41. What percentage of employees complete the recommended steps in development plans?
<--- Score

42. What are the most widely utilized legal plan benefits?
<--- Score

43. Are benefit plans reviewed annually?
<--- Score

44. Is a response plan in place for when the input, process, or output measures indicate an 'out-of-control' condition?
<--- Score

45. Is a response plan established and deployed?
<--- Score

46. Did the action plan accomplish what your organization needed?
<--- Score

47. Is there a documented and implemented monitoring plan?
<--- Score

48. How does the system you want support plans?
<--- Score

49. Are documented procedures clear and easy to follow for the operators?
<--- Score

50. What are the top performing learning organizations doing differently?
<--- Score

51. What key inputs and outputs are being measured on an ongoing basis?
<--- Score

52. Are compensation levels/plans monitored and reviewed annually?
<--- Score

53. What sequence of topics is most effective for a specific learner?
<--- Score

54. Has your organization assigned responsibilities and provided role clarity in carrying out the plan?
<--- Score

55. Are your stakeholder engagement tactics well planned and well executed?
<--- Score

56. Which plans have the best value in benefits?
<--- Score

57. How will report readings be checked to effectively monitor performance?
<--- Score

58. Does the system offset advanced sick leave balances against subsequent pay period accruals of sick leave?
<--- Score

59. How many records of historical learner completion records or data will you want loaded?
<--- Score

60. Do you have data of some size and scale?
<--- Score

61. What kinds of information are collected as a function of the monitoring of individuals?
<--- Score

62. How will input, process, and output variables be checked to detect for sub-optimal conditions?
<--- Score

63. Does the system process earnings adjustments on a begin date/end date basis?
<--- Score

64. Are you hiring according to plan?
<--- Score

65. Are the planning assumptions defined?
<--- Score

66. Who is the HR Analytics process owner?
<--- Score

67. How many currencies are being planned?
<--- Score

68. Have new or revised work instructions resulted?
<--- Score

69. Has the improved process and its steps been standardized?
<--- Score

70. Is the compensation plan communicated to all employees?
<--- Score

71. Does the HR Analytics performance meet the customer's requirements?
<--- Score

72. Is there a control plan in place for sustaining improvements (short and long-term)?
<--- Score

73. What external vendors do you plan to use for job posting and background screening for new hires?
<--- Score

74. Is there a transfer of ownership and knowledge to process owner and process team tasked with the responsibilities.
<--- Score

75. Are new process steps, standards, and documentation ingrained into normal operations?
<--- Score

76. How many job requisitions do you plan on converting at go live?
<--- Score

77. Do you develop your business plans differently?
<--- Score

78. What is the agility of your workforce to scale up in a large scale emergency situation?
<--- Score

79. What learner actions are associated with more learning?
<--- Score

80. How will new or emerging customer needs/requirements be checked/communicated to orient the process toward meeting the new specifications and continually reducing variation?
<--- Score

81. How do you increase your workforce planning and analytics maturity?
<--- Score

82. Have the ability to learn and master all aspects of the job once hired?
<--- Score

83. Have your organizations strategies on which the plan is based changed?
<--- Score

84. What is it that a legal plan should measure when it measures an abandonment rate?
<--- Score

85. How many non accruing time off plans do you

need configured?
<--- Score

86. Are there any needed adjustments to the plan and strategies?
<--- Score

87. How do you plan a large workforce from scratch?
<--- Score

88. When does your planning process typically begin?
<--- Score

89. Does the system determine and apply the appropriate pay scale for the employee?
<--- Score

90. What would it means in short and long term organizational planning?
<--- Score

91. Does job training on the documented procedures need to be part of the process team's education and training?
<--- Score

92. What other areas of the group might benefit from the HR Analytics team's improvements, knowledge, and learning?
<--- Score

93. How many learning instructors will need to be loaded?
<--- Score

94. Will any special training be provided for results interpretation?
<--- Score

95. Is reporting being used or needed?
<--- Score

96. How do you learn about the machinery?
<--- Score

97. Which benefits will employees with an acute need for legal plan help regard as extremely important?
<--- Score

98. Are planning assumptions still valid?
<--- Score

Add up total points for this section:
_____ = Total points for this section

Divided by: ______ (number of statements answered) = ______ Average score for this section

Transfer your score to the HR Analytics Index at the beginning of the Self-Assessment.

CRITERION #7: SUSTAIN:

INTENT: Retain the benefits.

In my belief, the answer to this question is clearly defined:

5 Strongly Agree

4 Agree

3 Neutral

2 Disagree

1 Strongly Disagree

1. Are return-to-work programs checked for effectiveness?
<--- Score

2. Who are the top talent and critical employees?
<--- Score

3. How many leave types do you have?
<--- Score

4. How many employees in total are being paid via

payroll?
<--- Score

5. What is the likelihood of attracting appropriately skilled talent?
<--- Score

6. Are risk management tasks balanced centrally and locally?
<--- Score

7. How well is your strategic work currently being performed?
<--- Score

8. Do lateral moves make employees promotable?
<--- Score

9. Are procedures documented for managing HR Analytics risks?
<--- Score

10. What is your voluntary turnover rate among all individuals in hard to hire positions?
<--- Score

11. Are you sure your organization is aware about the relation of employee mixture?
<--- Score

12. What alternative responses are available to manage risk?
<--- Score

13. Does the system accrue leave for part time employees?

<--- Score

14. How would you suggest your organization better prepare for that change?
<--- Score

15. What types of people tend to do well here?
<--- Score

16. Is the HR Analytics solution sustainable?
<--- Score

17. What knowledge or experience is required?
<--- Score

18. Does hris select the right persons to be trained at the right time?
<--- Score

19. How will knowledge make it to replacements?
<--- Score

20. What does your training fee cover?
<--- Score

21. What percentage of your operating expenses is invested in compensation and benefits?
<--- Score

22. What are the strengths of your organization related to your workforce?
<--- Score

23. Do you have the right leadership in place?
<--- Score

24. Does the existing structure correspond to your organizations function?
<--- Score

25. How do you manage freelance and outsourced staff?
<--- Score

26. Is the work to date meeting requirements?
<--- Score

27. Is there any other HR Analytics solution?
<--- Score

28. What percentage of new hires are terminated in the first six months?
<--- Score

29. Where do you find corresponding quality candidates?
<--- Score

30. Is there training of managers and employees about personnel files?
<--- Score

31. How do you reduce absenteeism and balance permanent versus contract staff?
<--- Score

32. What are other organizations looking for?
<--- Score

33. What talent dimensions predict success now and in the future in your organization?
<--- Score

34. What kinds of senior management personalities and work styles currently exist in your organization?
<--- Score

35. How do you more effectively correlate benefits and related investments in wellness with productivity?
<--- Score

36. Who manages supplier risk management in your organization?
<--- Score

37. Do you collect feedback on your employees?
<--- Score

38. What percentage of your workforce moves internally each year between lines of business or functional areas?
<--- Score

39. How many locations does the company have?
<--- Score

40. Has your organization allocated the necessary resources to ensure an appropriate response?
<--- Score

41. What is the geographic and functional scope?
<--- Score

42. How successful have transformational change programs been in the past?
<--- Score

43. How does the employee compare to others applying for the same job?
<--- Score

44. What are the benefits and payoff?
<--- Score

45. Which best describes your role in your organization?
<--- Score

46. How many employees do you really have?
<--- Score

47. Does management have the right priorities among projects?
<--- Score

48. Are disciplinary actions for violating workplace policies flexible?
<--- Score

49. What are the key capabilities that your organization might consider?
<--- Score

50. What is your revenue per employee?
<--- Score

51. Can any human do human resource management?
<--- Score

52. Do current labor agreements contain barriers to increased productivity, increased work

flexibility, and increased use of contractors?
<--- Score

53. How does your organization gets a return on its investment?
<--- Score

54. Are employees aware of their status?
<--- Score

55. What kind of management support do you have for kind of initiatives?
<--- Score

56. What percentage of your employment offers are accepted?
<--- Score

57. What training would be most helpful to your future?
<--- Score

58. Which department do you belong to?
<--- Score

59. What types of roles have the highest attrition rate?
<--- Score

60. What organizational unit should be the target of the capability audit?
<--- Score

61. Are the onboarding forms capturing qualifications and skills of new hires in a systematic way?

<--- Score

62. How much change management is enough?
<--- Score

63. What types of personalities and work styles might represent too much change for your organization to absorb well?
<--- Score

64. What information about your workforce would help you run your business better?
<--- Score

65. Is it organizations responsibility to nudge employees towards good behavior?
<--- Score

66. How would you rate the IT efficiency in your department?
<--- Score

67. Can you integrate quality management and risk management?
<--- Score

68. How does human resource management influence organizational outcomes?
<--- Score

69. Do the workload and workforce gaps still exist?
<--- Score

70. Do your current pay programs support retention of the best talent?
<--- Score

71. How many hours a week on average do you expect senior management to work?
<--- Score

72. Are you hiring, promoting and retaining the best talent?
<--- Score

73. Have you considered bottom line predictors?
<--- Score

74. Where does the model predict retention and employee quits?
<--- Score

75. Have organization strategies changed?
<--- Score

76. What HR Analytics data should be managed?
<--- Score

77. Does best practice HRM only work for intrinsically motivated employees?
<--- Score

78. Are current employees given appropriate consideration for promotion or lateral position changes?
<--- Score

79. What business asks post training?
<--- Score

80. Who has formal and informal influence on the program?

<--- Score

81. Which staff will be accountable for producing insights?
<--- Score

82. How should workforce equity be approached?
<--- Score

83. What differentiates you in the marketplace?
<--- Score

84. How is workforce equity operationalized?
<--- Score

85. What is your current position within your supply chain?
<--- Score

86. What happens to the supervisors when turnover is so high?
<--- Score

87. Do adequate staffing levels exist?
<--- Score

88. What is the future of peer review?
<--- Score

89. How critical is the programs mission or purpose to your organizations mission as a whole?
<--- Score

90. Are the most efficient solutions problem-specific?
<--- Score

91. What factors are driving that change?
<--- Score

92. What is the theory of the business?
<--- Score

93. What is the average time between promotions?
<--- Score

94. How can work be simplified, making systems easy to use?
<--- Score

95. Does the system enable on line approval or disapproval of recurring actions?
<--- Score

96. How does compensation & benefits policies affect attrition rate of your organization?
<--- Score

97. Is there any way to speed up the process?
<--- Score

98. How can total rewards professionals apply concepts to everyday business practices?
<--- Score

99. What are the critical workforce segments?
<--- Score

100. Do your new hires continually increase the overall quality and productivity of your workforce?
<--- Score

101. Where should you hire the top talent from?
<--- Score

102. Will the new system increase or decrease dependence on IT resources?
<--- Score

103. Does your business afford a bad hire?
<--- Score

104. What are the HR Analytics design outputs?
<--- Score

105. How will your organization manage the turnover of key positions every few years as opposed to every decade?
<--- Score

106. What can the information be used for?
<--- Score

107. How does the skills profile of your workforce compare to your competitors?
<--- Score

108. How do you want the consumers of your business to view your organization?
<--- Score

109. Did organization get a return on investment?
<--- Score

110. Are there certain types of jobs that may be hard to fill?
<--- Score

111. What procedures are used for hiring in your organization?
<--- Score

112. Do you build or buy your workforce?
<--- Score

113. Where is your organization heading?
<--- Score

114. Why are there high levels of manager dissatisfaction?
<--- Score

115. Is your talent management strategy aligned with your organizations mission and goals?
<--- Score

116. How is headcount allocated in terms of broad departmental activities or duties?
<--- Score

117. What roles ranked highest in job satisfaction?
<--- Score

118. Are you leveraging the most relevant and appropriate talent management practices?
<--- Score

119. How do you accelerate the skills of your emerging talent in strategic roles?
<--- Score

120. Have the right attitude towards work?
<--- Score

121. Do you have a strategy for sharing information to downstream internal and external systems?
<--- Score

122. What is industrial/organizational psychology?
<--- Score

123. How must the information be protected?
<--- Score

124. How many part time, temporary, or contract employees exist across your organization?
<--- Score

125. How long does it take you to respond to a new request for information?
<--- Score

126. Do you have the right coaches and leaders grooming your rising executives?
<--- Score

127. Where are resource constraints being felt most acutely?
<--- Score

128. What is driving the urgency around performance management?
<--- Score

129. When can an HRIS be said to be successful?
<--- Score

130. Are your employees engaged within your

organization through the use of surveys?
<--- Score

131. What percentage of total compensation is direct compensation versus benefits?
<--- Score

132. Has the deal between your organization and its employees changed?
<--- Score

133. What users will be impacted?
<--- Score

134. What is the key to transformation that works?
<--- Score

135. What percentage of new hires leave voluntarily in the first six months?
<--- Score

136. What is the setting in which you work?
<--- Score

137. Why is time to value so crucial to success?
<--- Score

138. Which skills should you have internally versus contract with outside providers?
<--- Score

139. Where will the information come from?
<--- Score

140. Are you good at the things which matter most?

<--- Score

141. Did organization get a return on its investment?
<--- Score

142. What is the evidence on evidence based management?
<--- Score

143. Is the mission clear and communicated, understood and embraced by employees?
<--- Score

144. What are your organizations hiring and turnover rates?
<--- Score

145. What metrics does your organization value most?
<--- Score

146. Do you know your organizations workforce demographics?
<--- Score

147. What are your business objectives?
<--- Score

148. Which departments are producing your best candidates?
<--- Score

149. How do you deliver the best possible offer or interaction at the next point of customer contact?
<--- Score

150. How many people are performing each job?
<--- Score

151. Do you let your mind wander during business considerations or executive meetings?
<--- Score

152. Why should your employer support it?
<--- Score

153. How would you rate the communications between IT groups in other departments?
<--- Score

154. How do you build systems to sustain the change?
<--- Score

155. Are the risks fully understood, reasonable and manageable?
<--- Score

156. Why do you find it difficult to hire talent with certain skill sets?
<--- Score

157. What systems/processes must you excel at?
<--- Score

158. Where and how does your culture evolve?
<--- Score

159. How do you fill the skills gap in your workforce?
<--- Score

160. How strong is the correlation between engagement and turnover?
<--- Score

161. How does the new work environment differ?
<--- Score

162. Who are the HR Analytics decision makers?
<--- Score

163. What percentage of maintenance work is performed on budget and on schedule?
<--- Score

164. Why cultivate an engaged workforce?
<--- Score

165. What account types are budgeted?
<--- Score

166. What percentage of your first choice candidates accept your offers?
<--- Score

167. What tools and models are used to project distribution line replacement?
<--- Score

168. When are your trainings scheduled?
<--- Score

169. How many people have left your organization?
<--- Score

170. What are the critical jobs to the overall success of your organization?
<--- Score

171. What are your organizational strengths?
<--- Score

172. What is your inventory of talent in your strategic positions?
<--- Score

173. What is your organization as a system?
<--- Score

174. What should you be applying for business transformation?
<--- Score

175. What is coaching and how could it add value to your organization?
<--- Score

176. How would you rate how well IT staff work together between IT departments?
<--- Score

177. What percentage of your workforce is promoted annually?
<--- Score

178. Does the vendor provide suitable training for users?
<--- Score

179. How do you become your organization person first and foremost?

<--- Score

180. How can it be assured that the project is successfully delivered without any defect?
<--- Score

181. Are you in a position to move ahead of your competitors?
<--- Score

182. Who will replace your current leaders and when?
<--- Score

183. What to think of the negative consequences that might occur when employees become mini companies?
<--- Score

184. Are jobs becoming more precarious?
<--- Score

185. What do you believe are the biggest people challenges created by your organizations business strategy?
<--- Score

186. What is the future role of talent acquisition?
<--- Score

187. Does the system accommodate requests for waiver of collections for leave advances?
<--- Score

188. Is there a commitment and budget to grow talent with internal staff?

<--- Score

189. Is it possible for one system to do everything?
<--- Score

190. What is HR Analytics risk?
<--- Score

191. Do you retain your key/most productive people at a higher rate than your best competitors do?
<--- Score

192. How can the system help raise awareness among provider workforces of performance relative to peers, linking quality and other outcomes to how well run your organization is?
<--- Score

193. Did you accomplish all of your personal and professional goals?
<--- Score

194. How efficient/ productive is recruiting function?
<--- Score

195. What percentage of managers send employees to training sessions?
<--- Score

196. What is the ratio of customer facing or revenue generating staff to back office staff?
<--- Score

197. What key talent are you losing to external

companies?
<--- Score

198. How do you tap into your employees connections to augment your recruiting efforts?
<--- Score

199. Do you have a culture of inclusion?
<--- Score

200. What % of your workforce is aligned to losses?
<--- Score

201. What is your organization good at doing and what should it be known for?
<--- Score

202. Are employees informed/oriented regarding their benefits?
<--- Score

203. What systems does your cybersecurity strategy include?
<--- Score

204. Does the system track employee relations actions from initiation through completion?
<--- Score

205. Are the demands for value and service increasing?
<--- Score

206. What discoveries did you make about yourself, your team, and your world?
<--- Score

207. What percentage of underperforming managers turn around performance by the next review period?
<--- Score

208. Why should customers buy from your organization, as opposed to one of your competitors?
<--- Score

209. Is your performance management system a tool to execute business strategy?
<--- Score

210. Are performance reviews being completed?
<--- Score

211. What is the style of talent you are attracting?
<--- Score

212. What are you doing every single day to emphasize your leadership of the entire organization?
<--- Score

213. Do you have the optimal project management team structure?
<--- Score

214. How do you ensure that action will take place?
<--- Score

215. Is the effort still involve departments?
<--- Score

216. Are your structures aligned with internal and external values around the world?
<--- Score

217. What operating models and organization structures can help you manage growth in the new normal?
<--- Score

218. Are you achieving appropriate value from in sourced or out sourced providers?
<--- Score

219. What work would you choose to do?
<--- Score

220. How do you determine which employees are candidates for succession?
<--- Score

221. What is currently happening or has happened related to your workforce and why?
<--- Score

222. What are other organizations to do?
<--- Score

223. Why do employees choose to leave your organization?
<--- Score

224. Do you know what is already in place?
<--- Score

225. Do you really know what your employees

think?
<--- Score

226. What is your current talent landscape?
<--- Score

227. Has your organization recruiting and hiring strategy supported short and long term goals?
<--- Score

228. What kinds of reports can be produced on individuals?
<--- Score

229. What are the average sales of your most engaged units?
<--- Score

230. Does the system generate payment of unpaid compensation to beneficiaries?
<--- Score

231. What do you do about a skills shortage in your organizations?
<--- Score

232. Where do you add value directly to the customer experience versus indirectly through employees?
<--- Score

233. What percentage of your leadership team is from a protected group?
<--- Score

234. How many passes did the white team make?

<--- Score

235. Where model predicts attrition and employee stays?
<--- Score

236. What are you going information to do with it?
<--- Score

237. How often does your board consider and strategize talent management?
<--- Score

238. How many dollars of operating profit are generated for every dollar invested in compensation and benefits?
<--- Score

239. Which lines of business or managers have the highest voluntary turnover rates?
<--- Score

240. How do your emotions shape your experience and sustain your desired routines?
<--- Score

241. Are users satisfied with its use and benefits?
<--- Score

242. How much more should a top performer get than an average performer?
<--- Score

243. Are you adequately rewarding your value creators for success?
<--- Score

244. Should you still have secret talent reviews and performance ratings?
<--- Score

245. How do you Value Talent Management?
<--- Score

246. How efficient/productive is recruiting function?
<--- Score

247. How does your organization compare to competitors with respect to diversity?
<--- Score

248. What does the competency model of a leader in the future of work look like?
<--- Score

249. How does it fit with your business?
<--- Score

250. Is the workplace environment maintained with safety in mind?
<--- Score

251. How many hours do you work per week?
<--- Score

252. What should you do based on the insights available?
<--- Score

253. What matters to your organization?
<--- Score

254. Are exempt employees accurately classified?
<--- Score

255. What is your time to fill by job grade and geographic location?
<--- Score

256. What does the business services sector look like?
<--- Score

257. Are the members, and especially you the leader, clear on who is sponsoring the project?
<--- Score

258. What percentage of your employees are fully engaged and are doing best work at all times?
<--- Score

259. Who are the HR Analytics decision-makers?
<--- Score

260. How often do you actually stop to think about how to challenge old ways of doing things?
<--- Score

261. Does your organization perform customer research?
<--- Score

262. Do you have the right technical team?
<--- Score

263. What do organizations do differently when it comes to technology?

<--- Score

264. Do you have the right people to achieve your short term and long term business goals?
<--- Score

265. Do you know whether existing staff will transfer across to the external provider?
<--- Score

266. What talent groups or categories will be critically important to achieving your strategic goals?
<--- Score

267. Does your client have multiple public cloud services?
<--- Score

268. How do you delight your internal and external customers?
<--- Score

269. Do you have the tools and techniques?
<--- Score

270. How do you bind employees more to your organization?
<--- Score

271. How well positioned is your organization to respond to the effects of future trends?
<--- Score

272. What are the characteristics of anticipated jobs?

<--- Score

273. How do you give someone a pay rise or a promotion in a way that maximizes happiness?
<--- Score

274. Does the current workforce have the adequate competencies to efficiently accomplish objectives?
<--- Score

275. Does your organization want to become fast and nimble to market?
<--- Score

276. Do benefits that have a high usage have more value to employees?
<--- Score

277. Where in your organization are your high performers concentrated?
<--- Score

278. Is there a correlation in your organization between spending on training and your organizations profitability?
<--- Score

279. What percentage of your current managerial positions are occupied by specific diversity groups?
<--- Score

280. Does the system calculate indebtedness for leave advances upon separation?
<--- Score

281. What is the key to finding the right BI tools?
<--- Score

282. Where is your business already using AI?
<--- Score

283. What are the leading industries and companies?
<--- Score

284. Does the system finalize personnel actions?
<--- Score

285. How do you grow your employer brand?
<--- Score

286. Which recruiting sources provide the highest performing employees?
<--- Score

287. What constitutes success in the eyes of the CEO?
<--- Score

288. How do people in your organization prefer to work together employee to employee?
<--- Score

289. Does the system enable on line approval or disapproval of corresponding recurring actions?
<--- Score

290. Do systems articulate into other HRM activities?
<--- Score

291. Is your performance unusual for your industry?
<--- Score

292. Does anyone really know the specifics of what a key employee is doing?
<--- Score

293. How should compensation be structured?
<--- Score

294. How are you aligning your talent accordingly?
<--- Score

295. How will you ensure that all the promised benefits are achieved?
<--- Score

296. What is the timeline for system implementation?
<--- Score

297. Are departmental goals and objectives clear, measurable, and realistic?
<--- Score

298. How well do you really know your organization?
<--- Score

299. What things affect your health maybe things that are missing in your community?
<--- Score

300. Does the system track completion of training

by employees?
<--- Score

301. Are all employees equally considered for leadership positions?
<--- Score

302. Does the system generate the proper type of information?
<--- Score

303. How well does your talent in strategic roles compare to the market supply?
<--- Score

304. Do you currently use goals and competencies?
<--- Score

305. Are performance appraisals completed and on time?
<--- Score

306. What is holding organizations back?
<--- Score

307. Do you have the skills to compete?
<--- Score

308. How is your organization perceived in the eyes of todays candidates?
<--- Score

309. What are the weaknesses of your organization related to your workforce?
<--- Score

310. Do you really pay for performance?
<--- Score

311. Are you achieving value from existing providers?
<--- Score

312. Is easy access to information part of your workforce investment?
<--- Score

313. How much do high performance work practices matter?
<--- Score

314. Is the position difficult to recruit for, or is that talent hard to find?
<--- Score

315. Which specific factors increases or decreases the probability of attrition?
<--- Score

316. Did you meet your business outcomes?
<--- Score

317. Do the viable solutions scale to future needs?
<--- Score

318. How much of your employee turnover consists of regretted loss?
<--- Score

319. Have you ever considered how your employees behaviours affect your organizations

overall success?
<--- Score

320. What is the average absenteeism rate?
<--- Score

321. Do you manage or supervise other employees?
<--- Score

322. How do you help your organization?
<--- Score

323. How innovative is your organization culture?
<--- Score

324. What percent of trained individuals are observed to use a specific skill or knowledge on the job?
<--- Score

325. Which roles may become redundant in the future?
<--- Score

326. What are the talent insights critical to organizations?
<--- Score

327. Who will facilitate the team and process?
<--- Score

328. What type of daily requests and which employee transactions would make the most sense to automate?
<--- Score

329. What might happen in the future?
<--- Score

330. What are your mission, vision, and values?
<--- Score

331. How differently do you pay your top contributors from your average contributors?
<--- Score

332. Is it time for a director of extended workforce management or a dedicated workforce management office?
<--- Score

333. Are you ready to adopt shared services?
<--- Score

334. What categories of individuals are covered in the system?
<--- Score

335. What are the implications for the ways you manage your people?
<--- Score

336. Which employees are your competition taking?
<--- Score

337. Do employees perform effectively in new roles in key positions?
<--- Score

338. Why does workforce equity matter?

<--- Score

339. Did you accurately capture how work was done?
<--- Score

340. How strong is your employer brand compared to your competitors?
<--- Score

341. How many inquiries or applications do you receive per open position?
<--- Score

342. How will divisions, work units, and jobs be designed?
<--- Score

343. How is the work currently being done and by whom?
<--- Score

344. Is your organization healthy enough to retain the new talent?
<--- Score

345. What suggestions or feedback do you share that would make your organization stronger and more successful?
<--- Score

346. Which personnel association uses the idea of bundles of integral HRM practices?
<--- Score

347. How do movements compare to local

unemployment rates over time?
<--- Score

348. How will work flow into each part of your organization?
<--- Score

349. What is the level of unemployment?
<--- Score

350. How long to keep data and how to manage retention costs?
<--- Score

351. What is your ratio of managerial to non managerial staff?
<--- Score

352. Are there certain occupational groups with increasing worker turnover?
<--- Score

353. How long can an employee be temporary?
<--- Score

354. Do you feel there is enough information given on the benefits?
<--- Score

355. What is your voluntary turnover rate among top performers?
<--- Score

356. Are infrastructure and security in place?
<--- Score

357. Does your organization complete the investigations within the applicable prescribed time frame?
<--- Score

358. Do you physically handle the job?
<--- Score

359. Are total compensation letters provided?
<--- Score

360. Did you handle your complaint/grievance in an appropriate and timely way?
<--- Score

361. What is the average time to full productivity for new hires?
<--- Score

362. What competencies currently exist in your organization?
<--- Score

363. Which operating platform does your organization currently use?
<--- Score

364. Are the key business and technology risks being managed?
<--- Score

365. What is your employment brand strength among prospective employees?
<--- Score

366. Why are other organizations stuck here?

<--- Score

367. Is your leadership team effective?
<--- Score

368. How do you infuse local market and social insights into product merchandising and pricing?
<--- Score

369. What is the internal mobility and experience across functions of your customer facing roles?
<--- Score

370. What is human capital or human resources management for?
<--- Score

371. What is your talent bench strength?
<--- Score

372. What are the reporting relationships of jobs?
<--- Score

373. Is HR Analytics documentation maintained?
<--- Score

374. Does the target have multiple HRIS or payroll systems?
<--- Score

375. When should a process be art not science?
<--- Score

376. What assumptions are made about the solution and approach?
<--- Score

377. How satisfied are employees who work in the manufacturing department?
<--- Score

378. How does the performance of your new hiresvary by recruiter or hiring manager?
<--- Score

379. Does your current workforce have the capacity and skills to meet production demands?
<--- Score

380. Are you maintaining a diverse workforce?
<--- Score

381. What is the best way to close the gap?
<--- Score

382. Has turnover reduced the competencies of a certain occupational group?
<--- Score

383. What are the processes for audit reporting and management?
<--- Score

384. What are the potential benefits?
<--- Score

385. What kinds of products or services would your customers value in each scenario?
<--- Score

386. How will you know whether the change has been successful?

<--- Score

387. Are the productivity and utilization level of the workforces appropriate?
<--- Score

388. Is there a mature market for external provision of services?
<--- Score

389. What are your role and responsibilities in your organization?
<--- Score

390. Does a positive gain spiral of resources, efficacy beliefs and engagement exist?
<--- Score

391. What percentage of total compensation is tied to performance?
<--- Score

392. How will corresponding data be collected?
<--- Score

393. How will the data be checked for quality?
<--- Score

394. What good is the best functionality if only a select few in your organization can access it?
<--- Score

395. Does your organization have a referral program?
<--- Score

396. Have you given enough time for members to try out and become comfortable with the new technologies?
<--- Score

397. How do investments in your employees actually affect workforce performance?
<--- Score

398. Are you hiring high performance employees?
<--- Score

399. How well do you know your workforce?
<--- Score

400. What is your voluntary turnover rate among key executives?
<--- Score

401. Has your stock options program added to or diluted your shares value?
<--- Score

402. What competencies, incentives and work practices support high performance?
<--- Score

403. What is the human capital value creation by different job roles?
<--- Score

404. How high is your annual employee turnover?
<--- Score

405. When was the last time you did something seriously creative?

<--- Score

406. How many members/trainees will be there in each batch?
<--- Score

407. How does hrm support sustainable business performance and growth?
<--- Score

408. What is the HR Analytics business impact?
<--- Score

409. How accurate are your pre employment tests in predicting the quality of a hire?
<--- Score

410. Which functions of HRM deals with Collective Bargaining?
<--- Score

411. What approach will provide the greatest probability of business success?
<--- Score

412. Where does your business element want to be?
<--- Score

413. Are the supply portfolio principles, goals, and objectives for mass market default customers reasonable and appropriate to ensure continuity of supply?
<--- Score

414. Are your organization and staffing of

forecasting functions reasonable?
<--- Score

415. What types of culture change are you looking for?
<--- Score

416. How is the information kept accurate?
<--- Score

417. How efficient/productive is your recruiting function?
<--- Score

418. How do you capture value, now and in the future, for your chosen customers?
<--- Score

419. Do you have the talent for your organization now and in the future?
<--- Score

420. Which models are applicable for your business?
<--- Score

421. Does hrm facilitate employee creativity and organizational innovation?
<--- Score

422. Did it take a systematic stakeholder perspective?
<--- Score

423. Do employee behaviors link to the culture?
<--- Score

424. What is the level of budget detail the Board sees and what are responsibilities with regard to the budgets?
<--- Score

425. Do any key positions have no successors?
<--- Score

426. What are the recruitment channels that your top performers are picked from?
<--- Score

427. What are the appropriate delivery models for your organization?
<--- Score

428. What skills will you have in the future?
<--- Score

429. Are top performers differentiated?
<--- Score

430. Did it foster healthy employee employer relationships?
<--- Score

431. What HR Analytics data will be collected?
<--- Score

432. What information are people requesting?
<--- Score

433. Where do you find quality candidates?
<--- Score

434. What is your voluntary turnover rate among individuals in critical roles?
<--- Score

435. Does the responsible management official directly involved in the dispute have settlement authority?
<--- Score

436. Do senior management cultures affect performance?
<--- Score

437. What strategies should your organization use to retain competencies?
<--- Score

Add up total points for this section:
_____ = Total points for this section

Divided by: ______ (number of statements answered) = ______
Average score for this section

Transfer your score to the HR Analytics Index at the beginning of the Self-Assessment.

HR Analytics and Managing Projects, Criteria for Project Managers:

1.0 Initiating Process Group: HR Analytics

1. What input will you be required to provide the HR Analytics project team?

2. Did you use a contractor or vendor?

3. How well did you do?

4. Do you know the roles & responsibilities required for this HR Analytics project?

5. What technical work to do in each phase?

6. Measurable - are the targets measurable?

7. Do you know all the stakeholders impacted by the HR Analytics project and what needs are?

8. What are the inputs required to produce the deliverables?

9. During which stage of Risk planning are modeling techniques used to determine overall effects of risks on HR Analytics project objectives for high probability, high impact risks?

10. When are the deliverables to be generated in each phase?

11. What were the challenges that you encountered during the execution of a previous HR Analytics project that you would not want to repeat?

12. Who is behind the HR Analytics project?

13. What must be done?

14. How well did the chosen processes produce the expected results?

15. Which six sigma dmaic phase focuses on why and how defects and errors occur?

16. Are the changes in your HR Analytics project being formally requested, analyzed, and approved by the appropriate decision makers?

17. What are the constraints?

18. Are there resources to maintain and support the outcome of the HR Analytics project?

19. Have you evaluated the teams performance and asked for feedback?

20. Just how important is your work to the overall success of the HR Analytics project?

1.1 Project Charter: HR Analytics

21. For whom?

22. HR Analytics project background: what is the primary motivation for this HR Analytics project?

23. When do you use a HR Analytics project Charter?

24. How much?

25. Is it an improvement over existing products?

26. Does the HR Analytics project need to consider any special capacity or capability issues?

27. Why do you manage integration?

28. Customer: who are you doing the HR Analytics project for?

29. What are some examples of a business case?

30. What are the assigned resources?

31. Run it as as a startup?

32. Why use a HR Analytics project charter?

33. What is in it for you?

34. Where does all this information come from?

35. Assumptions and constraints: what assumptions

were made in defining the HR Analytics project?

36. What material?

37. What are the assumptions?

38. Are there special technology requirements?

39. What is the business need?

40. Review the general mission What system will be affected by the improvement efforts?

1.2 Stakeholder Register: HR Analytics

41. What opportunities exist to provide communications?

42. What is the power of the stakeholder?

43. How will reports be created?

44. Is your organization ready for change?

45. Who wants to talk about Security?

46. Who is managing stakeholder engagement?

47. How should employers make voices heard?

48. Who are the stakeholders?

49. What & Why?

50. How big is the gap?

51. How much influence do they have on the HR Analytics project?

52. What are the major HR Analytics project milestones requiring communications or providing communications opportunities?

1.3 Stakeholder Analysis Matrix: HR Analytics

53. Marketing - reach, distribution, awareness?

54. What are the opportunities for communication?

55. What do you need to appraise?

56. What could your organization improve?

57. If you can not fix it, how do you do it differently?

58. Processes, systems, it, communications?

59. Who has control over whom?

60. Continuity, supply chain robustness?

61. What should thwe organizations stakeholders avoid?

62. Is there evidence that demonstrates the impact of education on the HR Analytics projects outcomes?

63. How do they affect the HR Analytics project and its outcomes?

64. Who is most dependent on the resources at stake?

65. What is social & public accountability ?

66. Competitor intentions - various?

67. What mechanisms are proposed to monitor and measure HR Analytics project performance in terms of social development outcomes?

68. Usps (unique selling points)?

69. How can you counter negative efforts?

70. How to measure the achievement of the Outputs?

71. What is the relationship among stakeholders?

72. Innovative aspects?

2.0 Planning Process Group: HR Analytics

73. Who are the HR Analytics project stakeholders?

74. Why is it important to determine activity sequencing on HR Analytics projects?

75. Is the schedule for the set products being met?

76. To what extent has the intervention strategy been adapted to the areas of intervention in which it is being implemented?

77. Why do it HR Analytics projects fail?

78. First of all, should any action be taken?

79. How will you know you did it?

80. Just how important is your work to the overall success of the HR Analytics project?

81. To what extent are the participating departments coordinating with each other?

82. If you are late, will anybody notice?

83. HR Analytics project assessment; why did you do this HR Analytics project?

84. Is the HR Analytics project making progress in helping to achieve the set results?

85. Professionals want to know what is expected from them; what are the deliverables?

86. What good practices or successful experiences or transferable examples have been identified?

87. How many days can task X be late in starting without affecting the HR Analytics project completion date?

88. How do you integrate HR Analytics project Planning with the Iterative/Evolutionary SDLC?

89. In what ways can the governance of the HR Analytics project be improved so that it has greater likelihood of achieving future sustainability?

90. Did you read it correctly?

91. Does it make any difference if you are successful?

92. What are the different approaches to building the WBS?

2.1 Project Management Plan: HR Analytics

93. Are there any windfall benefits that would accrue to the HR Analytics project sponsor or other parties?

94. Are there any scope changes proposed for a previously authorized HR Analytics project?

95. Development trends and opportunities. What if the positive direction and vision of your organization causes expected trends to change?

96. Is the budget realistic?

97. Does the selected plan protect privacy?

98. Will you add a schedule and diagram?

99. Are cost risk analysis methods applied to develop contingencies for the estimated total HR Analytics project costs?

100. What worked well?

101. What does management expect of PMs?

102. Are calculations and results of analyzes essentially correct?

103. Are alternatives safe, functional, constructible, economical, reasonable and sustainable?

104. How well are you able to manage your risk?

105. Is mitigation authorized or recommended?

106. Who is the sponsor?

107. How do you organize the costs in the HR Analytics project management plan?

108. Is there anything you would now do differently on your HR Analytics project based on past experience?

109. What are the training needs?

2.2 Scope Management Plan: HR Analytics

110. Personnel with expertise?

111. Are cause and effect determined for risks when they occur?

112. Is there a Steering Committee in place?

113. Who is doing what for whom?

114. Are HR Analytics project team members committed fulltime?

115. Are milestone deliverables effectively tracked and compared to HR Analytics project plan?

116. Are tasks tracked by hours?

117. Where do scope management processes fit in?

118. Have all team members been part of identifying risks?

119. Are target dates established for each milestone deliverable?

120. Are changes in scope (deliverable commitments) agreed to by all affected groups & individuals?

121. Have reserves been created to address risks?

122. How difficult will it be to do specific activities on this HR Analytics project?

123. Without-plan conditions?

124. Are there any windfall benefits that would accrue to the HR Analytics project sponsor or other parties?

125. Is each item clearly and completely defined?

126. How do you know when you are finished?

127. Describe how the deliverables will be verified against the HR Analytics project scope. To whom will the deliverables be first presented for inspection and verification?

128. Are risk oriented checklists used during risk identification?

129. Is it possible to track all classes of HR Analytics project work (e.g. scheduled, un-scheduled, defect repair, etc.)?

2.3 Requirements Management Plan: HR Analytics

130. Do you understand the role that each stakeholder will play in the requirements process?

131. Will you use tracing to help understand the impact of a change in requirements?

132. In case of software development; Should you have a test for each code module?

133. Who will finally present the work or product(s) for acceptance?

134. Is requirements work dependent on any other specific HR Analytics project or non-HR Analytics project activities (e.g. funding, approvals, procurement)?

135. Do you have price sheets and a methodology for determining the total proposal cost?

136. Does the HR Analytics project have a Change Control process?

137. What cost metrics will be used?

138. How will the information be distributed?

139. Subject to change control?

140. Have stakeholders been instructed in the Change

Control process?

141. How often will the reporting occur?

142. Did you distinguish the scope of work the contractor(s) will be required to do?

143. How will the requirements become prioritized?

144. What is a problem?

145. Who will initially review the HR Analytics project work or products to ensure it meets the applicable acceptance criteria?

146. Is the system software (non-operating system) new to the IT HR Analytics project team?

147. What is the earliest finish date for this HR Analytics project if it is scheduled to start on ...?

148. What information regarding the HR Analytics project requirements will be reported?

149. What went right?

2.4 Requirements Documentation: HR Analytics

150. What are the potential disadvantages/ advantages?

151. How do you get the user to tell you what they want?

152. Where do system and software requirements come from, what are sources?

153. Validity. does the system provide the functions which best support the customers needs?

154. Consistency. are there any requirements conflicts?

155. If applicable; are there issues linked with the fact that this is an offshore HR Analytics project?

156. Are there any requirements conflicts?

157. How to document system requirements?

158. What is the risk associated with the technology?

159. Can the requirements be checked?

160. Is the origin of the requirement clearly stated?

161. How much testing do you need to do to prove that your system is safe?

162. How linear / iterative is your Requirements Gathering process (or will it be)?

163. What happens when requirements are wrong?

164. Are there legal issues?

165. Has requirements gathering uncovered information that would necessitate changes?

166. How will they be documented / shared?

167. What are the attributes of a customer?

168. Is new technology needed?

169. Can you check system requirements?

2.5 Requirements Traceability Matrix: HR Analytics

170. What percentage of HR Analytics projects are producing traceability matrices between requirements and other work products?

171. How small is small enough?

172. Is there a requirements traceability process in place?

173. Will you use a Requirements Traceability Matrix?

174. Why do you manage scope?

175. Do you have a clear understanding of all subcontracts in place?

176. Describe the process for approving requirements so they can be added to the traceability matrix and HR Analytics project work can be performed. Will the HR Analytics project requirements become approved in writing?

177. How will it affect the stakeholders personally in career?

178. How do you manage scope?

179. What is the WBS?

180. Why use a WBS?

181. What are the chronologies, contingencies, consequences, criteria?

2.6 Project Scope Statement: HR Analytics

182. Are there backup strategies for key members of the HR Analytics project?

183. Risks?

184. What is the most common tool for helping define the detail?

185. Is your organization structure appropriate for the HR Analytics projects size and complexity?

186. Which risks does the HR Analytics project focus on?

187. Identify how your team and you will create the HR Analytics project scope statement and the work breakdown structure (WBS). Document how you will create the HR Analytics project scope statement and WBS, and make sure you answer the following questions: In defining HR Analytics project scope and the WBS, will you and your HR Analytics project team be using methods defined by your organization, methods defined by the HR Analytics project management office (PMO), or other methods?

188. Are there specific processes you will use to evaluate and approve/reject changes?

189. Have the configuration management functions been assigned?

190. If there is an independent oversight contractor, have they signed off on the HR Analytics project Plan?

191. Does the scope statement still need some clarity?

192. Is the HR Analytics project manager qualified and experienced in HR Analytics project management?

193. Are the meetings set up to have assigned note takers that will add action/issues to the issue list?

194. Will tasks be marked complete only after QA has been successfully completed?

195. Will an issue form be in use?

196. Is the HR Analytics project sponsor function identified and defined?

197. Is the scope of your HR Analytics project well defined?

198. Why do you need to manage scope?

199. If you were to write a list of what should not be included in the scope statement, what are the things that you would recommend be described as out-of-scope?

200. What is change?

2.7 Assumption and Constraint Log: HR Analytics

201. Model-building: what data-analytic strategies are useful when building proportional-hazards models?

202. Security analysis has access to information that is sanitized?

203. What strengths do you have?

204. Have you eliminated all duplicative tasks or manual efforts, where appropriate?

205. Contradictory information between document sections?

206. Does the system design reflect the requirements?

207. Is the definition of the HR Analytics project scope clear; what needs to be accomplished?

208. How are new requirements or changes to requirements identified?

209. Have adequate resources been provided by management to ensure HR Analytics project success?

210. Are there procedures in place to effectively manage interdependencies with other HR Analytics projects / systems?

211. Are processes for release management of

new development from coding and unit testing, to integration testing, to training, and production defined and followed?

212. Violation trace: why ?

213. If appropriate, is the deliverable content consistent with current HR Analytics project documents and in compliance with the Document Management Plan?

214. Have all stakeholders been identified?

215. Is the process working, and people are not executing in compliance of the process?

216. What is positive about the current process?

217. Is this process still needed?

218. Are funding and staffing resource estimates sufficiently detailed and documented for use in planning and tracking the HR Analytics project?

219. Diagrams and tables are included to account for complex concepts and increase overall readability?

220. What does an audit system look like?

2.8 Work Breakdown Structure: HR Analytics

221. How many levels?

222. Why would you develop a Work Breakdown Structure?

223. Who has to do it?

224. When would you develop a Work Breakdown Structure?

225. Where does it take place?

226. Do you need another level?

227. How much detail?

228. How will you and your HR Analytics project team define the HR Analytics projects scope and work breakdown structure?

229. What is the probability of completing the HR Analytics project in less that xx days?

230. Is it a change in scope?

231. How far down?

232. How big is a work-package?

233. When do you stop?

234. Why is it useful?

235. Can you make it?

236. Is the work breakdown structure (wbs) defined and is the scope of the HR Analytics project clear with assigned deliverable owners?

237. When does it have to be done?

2.9 WBS Dictionary: HR Analytics

238. Does the accounting system provide a basis for auditing records of direct costs chargeable to the contract?

239. Incurrence of actual indirect costs in excess of budgets, by element of expense?

240. Identify potential or actual budget-based and time-based schedule variances?

241. How detailed should a HR Analytics project get?

242. Changes in the current direct and HR Analytics projected base?

243. Are the bases and rates for allocating costs from each indirect pool consistently applied?

244. Does the contractor use objective results, design reviews and tests to trace schedule performance?

245. Are all authorized tasks assigned to identified organizational elements?

246. Are overhead cost budgets (or HR Analytics projections) established on a facility-wide basis at least annually for the life of the contract?

247. Are indirect costs accumulated for comparison with the corresponding budgets?

248. Are the procedures for identifying indirect costs

to incurring organizations, indirect cost pools, and allocating the costs from the pools to the contracts formally documented?

249. Are work packages assigned to performing organizations?

250. Are management actions taken to reduce indirect costs when there are significant adverse variances?

251. Are records maintained to show full accountability for all material purchased for the contract, including the residual inventory?

252. What are you counting on?

253. Are overhead costs budgets established on a basis consistent with anticipated direct business base?

254. Are procedures in existence that control replanning of unopened work packages, and are corresponding procedures adhered to?

255. Are overhead cost budgets established for each organization which has authority to incur overhead costs?

256. Are estimates of costs at completion utilized in determining contract funding requirements and reporting them?

2.10 Schedule Management Plan: HR Analytics

257. Identify the amount of schedule variation that triggers a warning. What happens if a warning is triggered?

258. Is there anything planned that does not need to be here?

259. Are all resource assumptions documented?

260. Must the HR Analytics project be complete by a specified date?

261. Have all unresolved risks been documented?

262. Does the schedule have reasonable float?

263. Were HR Analytics project team members involved in detailed estimating and scheduling?

264. Why time management?

265. What is the estimated time to complete the HR Analytics project if status quo is maintained?

266. Are software metrics formally captured, analyzed and used as a basis for other HR Analytics project estimates?

267. HR Analytics project definition & scope?

268. Time for overtime?

269. Have the procedures for identifying budget variances been followed?

270. Are vendor contract reports, reviews and visits conducted periodically?

271. Goal: is the schedule feasible and at what cost?

272. Is an industry recognized mechanized support tool(s) being used for HR Analytics project scheduling & tracking?

273. Will rolling way planning be used?

274. Are decisions captured in a decisions log?

2.11 Activity List: HR Analytics

275. What did not go as well?

276. What is the probability the HR Analytics project can be completed in xx weeks?

277. Should you include sub-activities?

278. Where will it be performed?

279. How can the HR Analytics project be displayed graphically to better visualize the activities?

280. How much slack is available in the HR Analytics project?

281. Is infrastructure setup part of your HR Analytics project?

282. Can you determine the activity that must finish, before this activity can start?

283. What are the critical bottleneck activities?

284. When do the individual activities need to start and finish?

285. How will it be performed?

286. When will the work be performed?

287. What went wrong?

288. How do you determine the late start (LS) for each activity?

289. What will be performed?

290. Are the required resources available or need to be acquired?

291. Who will perform the work?

292. What went well?

293. How difficult will it be to do specific activities on this HR Analytics project?

294. How detailed should a HR Analytics project get?

2.12 Activity Attributes: HR Analytics

295. Are the required resources available?

296. Would you consider either of corresponding activities an outlier?

297. How difficult will it be to do specific activities on this HR Analytics project?

298. Do you feel very comfortable with your prediction?

299. What conclusions/generalizations can you draw from this?

300. Activity: what is Missing?

301. Were there other ways you could have organized the data to achieve similar results?

302. What activity do you think you should spend the most time on?

303. Have you identified the Activity Leveling Priority code value on each activity?

304. Does your organization of the data change its meaning?

305. What is missing?

306. How much activity detail is required?

307. Which method produces the more accurate cost assignment?

308. Why?

309. Resource is assigned to?

310. How else could the items be grouped?

311. How difficult will it be to complete specific activities on this HR Analytics project?

312. How many days do you need to complete the work scope with a limit of X number of resources?

2.13 Milestone List: HR Analytics

313. How will you get the word out to customers?

314. How soon can the activity finish?

315. Reliability of data, plan predictability?

316. What specific improvements did you make to the HR Analytics project proposal since the previous time?

317. Information and research?

318. How soon can the activity start?

319. Describe your organizations strengths and core competencies. What factors will make your organization succeed?

320. Timescales, deadlines and pressures?

321. Describe the industry you are in and the market growth opportunities. What is the market for your technology, product or service?

322. Sustaining internal capabilities?

323. Sustainable financial backing?

324. Loss of key staff?

325. Environmental effects?

326. Who will manage the HR Analytics project on a

day-to-day basis?

327. How will the milestone be verified?

328. What date will the task finish?

329. How difficult will it be to do specific activities on this HR Analytics project?

2.14 Network Diagram: HR Analytics

330. Where do schedules come from?

331. What are the tools?

332. What must be completed before an activity can be started?

333. What is the lowest cost to complete this HR Analytics project in xx weeks?

334. What activities must occur simultaneously with this activity?

335. What is the completion time?

336. What is your organizations history in doing similar activities?

337. What job or jobs precede it?

338. What can be done concurrently?

339. Why must you schedule milestones, such as reviews, throughout the HR Analytics project?

340. If a current contract exists, can you provide the vendor name, contract start, and contract expiration date?

341. What controls the start and finish of a job?

342. Will crashing x weeks return more in benefits

than it costs?

343. What to do and When?

344. Review the logical flow of the network diagram. Take a look at which activities you have first and then sequence the activities. Do they make sense?

345. Can you calculate the confidence level?

346. How confident can you be in your milestone dates and the delivery date?

347. How difficult will it be to do specific activities on this HR Analytics project?

348. What activity must be completed immediately before this activity can start?

349. What are the Major Administrative Issues?

2.15 Activity Resource Requirements: HR Analytics

350. When does monitoring begin?

351. Are there unresolved issues that need to be addressed?

352. Do you use tools like decomposition and rolling-wave planning to produce the activity list and other outputs?

353. How do you manage time?

354. What are constraints that you might find during the Human Resource Planning process?

355. Anything else?

356. Why do you do that?

357. Which logical relationship does the PDM use most often?

358. Organizational Applicability?

359. How do you handle petty cash?

360. What is the Work Plan Standard?

361. Other support in specific areas?

362. How many signatures do you require on a

check and does this match what is in your policy and procedures?

2.16 Resource Breakdown Structure: HR Analytics

363. What is the number one predictor of a groups productivity?

364. Which resources should be in the resource pool?

365. Who will use the system?

366. Any changes from stakeholders?

367. What is the purpose of assigning and documenting responsibility?

368. What defines a successful HR Analytics project?

369. How should the information be delivered?

370. Who delivers the information?

371. What is each stakeholders desired outcome for the HR Analytics project?

372. The list could probably go on, but, the thing that you would most like to know is, How long & How much?

373. What defines a successful HR Analytics project?

374. Why is this important?

375. Who is allowed to see what data about which

resources?

376. Who needs what information?

377. Goals for the HR Analytics project. What is each stakeholders desired outcome for the HR Analytics project?

378. How can this help you with team building?

379. Who will be used as a HR Analytics project team member?

380. Is predictive resource analysis being done?

2.17 Activity Duration Estimates: HR Analytics

381. What distinguishes one organization from another in this area?

382. What is the difference between using brainstorming and the Delphi technique for risk identification?

383. Are activity dependencies identified?

384. How can others help HR Analytics project managers understand your organizational context for HR Analytics projects?

385. How difficult will it be to complete specific activities on this HR Analytics project?

386. Are procedures defined for calculating cost estimates?

387. How much time is required to develop it?

388. Do procedures exist describing how the HR Analytics project scope will be managed?

389. Why do you need a good WBS to use HR Analytics project management software?

390. What are the nine areas of expertise?

391. Are contingency plans created to prepare for risk

events to occur?

392. Which does one need in order to complete schedule development?

393. See what went wrong?

394. Why do you think schedule issues often cause the most conflicts on HR Analytics projects?

395. Who has the PRIMARY responsibility to solve this problem?

396. What is pmp certification, and why do you think the number of people earning it has grown so much in the past ten years?

397. Is the work performed reviewed against contractual objectives?

398. Are procedures defined by which the HR Analytics project scope may be changed?

2.18 Duration Estimating Worksheet: HR Analytics

399. What work will be included in the HR Analytics project?

400. What questions do you have?

401. What utility impacts are there?

402. When does your organization expect to be able to complete it?

403. What is cost and HR Analytics project cost management?

404. Does the HR Analytics project provide innovative ways for stakeholders to overcome obstacles or deliver better outcomes?

405. For other activities, how much delay can be tolerated?

406. Small or large HR Analytics project?

407. Is a construction detail attached (to aid in explanation)?

408. Done before proceeding with this activity or what can be done concurrently?

409. How should ongoing costs be monitored to try to keep the HR Analytics project within budget?

410. Is this operation cost effective?

411. What is your role?

412. Value pocket identification & quantification what are value pockets?

413. Why estimate costs?

414. Can the HR Analytics project be constructed as planned?

415. How can the HR Analytics project be displayed graphically to better visualize the activities?

2.19 Project Schedule: HR Analytics

416. How can you minimize or control changes to HR Analytics project schedules?

417. Are you working on the right risks?

418. Is HR Analytics project work proceeding in accordance with the original HR Analytics project schedule?

419. How does a HR Analytics project get to be a year late ?

420. How do you manage HR Analytics project Risk?

421. How can you address that situation?

422. How can slack be negative?

423. How much slack is available in the HR Analytics project?

424. Did the HR Analytics project come in under budget?

425. Why do you think schedule issues often cause the most conflicts on HR Analytics projects?

426. Schedule/cost recovery?

427. Master HR Analytics project schedule?

428. Is the structure for tracking the HR Analytics

project schedule well defined and assigned to a specific individual?

429. What is risk management?

430. Is there a Schedule Management Plan that establishes the criteria and activities for developing, monitoring and controlling the HR Analytics project schedule?

431. Does the condition or event threaten the HR Analytics projects objectives in any ways?

432. Was the HR Analytics project schedule reviewed by all stakeholders and formally accepted?

2.20 Cost Management Plan: HR Analytics

433. Why do you manage cost?

434. Will the earned value reporting interface between time and cost management?

435. Are updated HR Analytics project time & resource estimates reasonable based on the current HR Analytics project stage?

436. Has the budget been baselined?

437. Have the key elements of a coherent HR Analytics project management strategy been established?

438. Has the business need been clearly defined?

439. Were HR Analytics project team members involved in the development of activity & task decomposition?

440. Cost management – how will the cost of changes be estimated and controlled?

441. Do HR Analytics project managers participating in the HR Analytics project know the HR Analytics projects true status first hand?

442. Schedule contingency – how will the schedule contingency be administrated?

443. Are meeting minutes captured and sent out after the meeting?

444. What does this mean to a cost or scheduler manager?

445. Technical and functional?

446. Is the quality assurance team identified?

447. Pareto diagrams, statistical sampling, flow charting or trend analysis used quality monitoring?

448. Was the scope definition used in task sequencing?

449. Are meeting objectives identified for each meeting?

450. Does the detailed work plan match the complexity of tasks with the capabilities of personnel?

451. What will be the split of responsibilities of progress measurement and controls among the owner, contractor, subcontractors, and vendors?

2.21 Activity Cost Estimates: HR Analytics

452. Certification of actual expenditures?

453. Based on your HR Analytics project communication management plan, what worked well?

454. Were you satisfied with the work?

455. What is the last item a HR Analytics project manager must do to finalize HR Analytics project close-out?

456. How Award?

457. Would you hire them again?

458. What is the HR Analytics projects sustainability strategy that will ensure HR Analytics project results will endure or be sustained?

459. Will you use any tools, such as HR Analytics project management software, to assist in capturing Earned Value metrics?

460. What areas does the group agree are the biggest success on the HR Analytics project?

461. What is the activity inventory?

462. Specific - is the objective clear in terms of what,

how, when, and where the situation will be changed?

463. What communication items need improvement?

464. Padding is bad and contingencies are good. what is the difference?

465. What is a HR Analytics project Management Plan?

466. In which phase of the acquisition process cycle does source qualifications reside?

467. How do you change activities?

468. What happens if you cannot produce the documentation for the single audit?

469. Where can you get activity reports?

470. How do you do activity recasts?

471. What are the audit requirements?

2.22 Cost Estimating Worksheet: HR Analytics

472. Identify the timeframe necessary to monitor progress and collect data to determine how the selected measure has changed?

473. Ask: are others positioned to know, are others credible, and will others cooperate?

474. Does the HR Analytics project provide innovative ways for stakeholders to overcome obstacles or deliver better outcomes?

475. How will the results be shared and to whom?

476. What is the purpose of estimating?

477. What additional HR Analytics project(s) could be initiated as a result of this HR Analytics project?

478. What is the estimated labor cost today based upon this information?

479. Is it feasible to establish a control group arrangement?

480. What costs are to be estimated?

481. Can a trend be established from historical performance data on the selected measure and are the criteria for using trend analysis or forecasting methods met?

482. Is the HR Analytics project responsive to community need?

483. What will others want?

484. What can be included?

485. Will the HR Analytics project collaborate with the local community and leverage resources?

486. What info is needed?

487. Who is best positioned to know and assist in identifying corresponding factors?

488. What happens to any remaining funds not used?

2.23 Cost Baseline: HR Analytics

489. How difficult will it be to do specific tasks on the HR Analytics project?

490. What threats might prevent you from getting there?

491. Has the documentation relating to operation and maintenance of the product(s) or service(s) been delivered to, and accepted by, operations management?

492. How do you manage cost?

493. Impact to environment?

494. Has the appropriate access to relevant data and analysis capability been granted?

495. HR Analytics project goals -should others be reconsidered?

496. Definition of done can be traced back to the definitions of what are you providing to the customer in terms of deliverables?

497. Has the HR Analytics project documentation been archived or otherwise disposed as described in the HR Analytics project communication plan?

498. On time?

499. Does the suggested change request represent a

desired enhancement to the products functionality?

500. Have you identified skills that are missing from your team?

501. Are you asking management for something as a result of this update?

502. Have all the product or service deliverables been accepted by the customer?

503. Does the suggested change request seem to represent a necessary enhancement to the product?

504. Are procedures defined by which the cost baseline may be changed?

505. Is there anything unique in this HR Analytics projects scope statement that will affect resources?

2.24 Quality Management Plan: HR Analytics

506. Do you keep back-up copies of any data?

507. What are your key performance measures/ indicators for tracking progress relative to your action plans?

508. Are best practices and metrics employed to identify issues, progress, performance, etc.?

509. Have adequate resources been provided by management to ensure HR Analytics project success?

510. After observing execution of process, is it in compliance with the documented Plan?

511. Are you following the quality standards?

512. Was trending evident between reviews?

513. Is there a Quality Management Plan?

514. How do you ensure that protocols are up to date?

515. What procedures are used to determine if you use, and the number of split, replicate or duplicate samples taken at a site?

516. What are the established criteria that sampling / testing data are compared against?

517. How do you decide what information to record?

518. How do senior leaders create your organizational focus on customers and other stakeholders?

519. How are calibration records kept?

520. What data do you gather/use/compile?

521. Who is responsible?

522. How do you ensure that your sampling methods and procedures meet your data quality objectives?

523. How do you decide what information needs to be recorded?

2.25 Quality Metrics: HR Analytics

524. How do you calculate such metrics?

525. What can manufacturing professionals do to ensure quality is seen as an integral part of the entire product lifecycle?

526. What level of statistical confidence do you use?

527. How can the effectiveness of each of the activities be measured?

528. How effective are your security tests?

529. What are your organizations expectations for its quality HR Analytics project?

530. Where is quality now?

531. The metrics–what is being considered?

532. Are interface issues coordinated?

533. Were quality attributes reported?

534. Are documents on hand to provide explanations of privacy and confidentiality?

535. Are applicable standards referenced and available?

536. Are quality metrics defined?

537. Has risk analysis been adequately reviewed?

538. How do you measure?

539. Have alternatives been defined in the event that failure occurs?

540. How do you know if everyone is trying to improve the right things?

541. Are there already quality metrics available that detect nonlinear embeddings and trends similar to the users perception?

2.26 Process Improvement Plan: HR Analytics

542. What actions are needed to address the problems and achieve the goals?

543. Who should prepare the process improvement action plan?

544. Does your process ensure quality?

545. Has a process guide to collect the data been developed?

546. What makes people good SPI coaches?

547. Have the supporting tools been developed or acquired?

548. How do you manage quality?

549. What personnel are the sponsors for that initiative?

550. Where do you want to be?

551. Management commitment at all levels?

552. Modeling current processes is great, and will you ever see a return on that investment?

553. Everyone agrees on what process improvement is, right?

554. Does explicit definition of the measures exist?

555. Are you meeting the quality standards?

556. What personnel are the coaches for your initiative?

557. Have the frequency of collection and the points in the process where measurements will be made been determined?

558. Where are you now?

559. Are you making progress on your improvement plan?

560. Where do you focus?

561. To elicit goal statements, do you ask a question such as, What do you want to achieve?

2.27 Responsibility Assignment Matrix: HR Analytics

562. HR Analytics projected economic escalation?

563. Is data disseminated to the contractors management timely, accurate, and usable?

564. What are the known stakeholder requirements?

565. What do you do when people do not respond?

566. How many people do you need?

567. Is every signing-off responsibility and every communicating responsibility critically necessary?

568. What is the justification?

569. Undistributed budgets, if any?

570. Are estimates of costs at completion generated in a rational, consistent manner?

571. What expertise is not available in your department?

572. Are records maintained to show how management reserves are used?

573. Evaluate the performance of operating organizations?

574. Will too many Signing-off responsibilities delay the completion of the activity/deliverable?

575. How do you manage human resources?

576. How cost benefit analysis?

577. Why cost benefit analysis?

2.28 Roles and Responsibilities: HR Analytics

578. Accountabilities: what are the roles and responsibilities of individual team members?

579. How well did the HR Analytics project Team understand the expectations of specific roles and responsibilities?

580. What are your major roles and responsibilities in the area of performance measurement and assessment?

581. Implementation of actions: Who are the responsible units?

582. Required skills, knowledge, experience?

583. Is feedback clearly communicated and non-judgmental?

584. Influence: what areas of organizational decision making are you able to influence when you do not have authority to make the final decision?

585. What areas would you highlight for changes or improvements?

586. To decide whether to use a quality measurement, ask how will you know when it is achieved?

587. What expectations were met?

588. Do the values and practices inherent in the culture of your organization foster or hinder the process?

589. Does the team have access to and ability to use data analysis tools?

590. What is working well?

591. What should you highlight for improvement?

592. What areas of supervision are challenging for you?

593. Who is involved?

594. What should you do now to prepare yourself for a promotion, increased responsibilities or a different job?

595. Was the expectation clearly communicated?

596. How is your work-life balance?

2.29 Human Resource Management Plan: HR Analytics

597. Was your organizations estimating methodology being used and followed?

598. What commitments have been made?

599. Do HR Analytics project managers participating in the HR Analytics project know the HR Analytics projects true status first hand?

600. Have adequate resources been provided by management to ensure HR Analytics project success?

601. Have lessons learned been conducted after each HR Analytics project release?

602. Are staff skills known and available for each task?

603. Is the assigned HR Analytics project manager a PMP (Certified HR Analytics project manager) and experienced?

604. Is there any form of automated support for Issues Management?

605. Are the appropriate IT resources adequate to meet planned commitments?

606. How well does your organization communicate?

607. Have all involved HR Analytics project

stakeholders and work groups committed to the HR Analytics project?

608. Where is your organization headed?

609. Is stakeholder involvement adequate?

610. Are people being developed to meet the challenges of the future?

611. Are change requests logged and managed?

2.30 Communications Management Plan: HR Analytics

612. Who needs to know and how much?

613. In your work, how much time is spent on stakeholder identification?

614. Who is the stakeholder?

615. Who is involved as you identify stakeholders?

616. What does the stakeholder need from the team?

617. How were corresponding initiatives successful?

618. Who were proponents/opponents?

619. What communications method?

620. What is the stakeholders level of authority?

621. Are the stakeholders getting the information others need, are others consulted, are concerns addressed?

622. Which stakeholders can influence others?

623. How will the person responsible for executing the communication item be notified?

624. Do you feel more overwhelmed by stakeholders?

625. Who to share with?

626. Is the stakeholder role recognized by your organization?

627. Are you constantly rushing from meeting to meeting?

628. Are stakeholders internal or external?

629. Do you then often overlook a key stakeholder or stakeholder group?

630. What are the interrelationships?

2.31 Risk Management Plan: HR Analytics

631. Who should be notified of the occurrence of each of the indicators?

632. Is this an issue, action item, question or a risk?

633. Why might it be late?

634. How will the HR Analytics project know if your organizations risk response actions were effective?

635. Is the process being followed?

636. Is the process supported by tools?

637. Is the customer willing to participate in reviews?

638. Does the customer understand the software process?

639. How would you suggest monitoring for risk transition indicators?

640. How is risk response planning performed?

641. Financial risk: can your organization afford to undertake the HR Analytics project?

642. What is the likelihood?

643. How can the process be made more effective or

less cumbersome (process improvements)?

644. Which is an input to the risk management process?

645. What is the probability the risk avoidance strategy will be successful?

646. Risk probability and impact: how will the probabilities and impacts of risk items be assessed?

647. Risks should be identified during which phase of HR Analytics project management life cycle?

648. Do requirements demand the use of new analysis, design, or testing methods?

649. How is risk monitoring performed?

2.32 Risk Register: HR Analytics

650. Can the likelihood and impact of failing to achieve corresponding recommendations and action plans be assessed?

651. Are there any knock-on effects/impact on any of the other areas?

652. What may happen or not go according to plan?

653. How are risks identified?

654. What are you going to do to limit the HR Analytics projects risk exposure due to the identified risks?

655. How are risks graded?

656. Have other controls and solutions been implemented in other services which could be applied as an alternative to additional funding?

657. How could corresponding Risk affect the HR Analytics project in terms of cost and schedule?

658. What is your current and future risk profile?

659. What should the audit role be in establishing a risk management process?

660. Who needs to know about this?

661. What are the assumptions and current status that

support the assessment of the risk?

662. Severity Prediction?

663. Having taken action, how did the responses effect change, and where is the HR Analytics project now?

664. How is a Community Risk Register created?

665. Cost/benefit – how much will the proposed mitigations cost and how does this cost compare with the potential cost of the risk event/situation should it occur?

666. Contingency actions - planned actions to reduce the immediate seriousness of the risk when it does occur. What should you do when?

667. Which key risks have ineffective responses or outstanding improvement actions?

668. What evidence do you have to justify the likelihood score of the risk (audit, incident report, claim, complaints, inspection, internal review)?

669. Schedule impact/severity estimated range (workdays) assume the event happens, what is the potential impact?

2.33 Probability and Impact Assessment: HR Analytics

670. Have you worked with the customer in the past?

671. How will the consumption pattern change?

672. Who will be in command to monitor and control the performance of the consortium members (consortium leader/client)?

673. Do you train all developers in the process?

674. Are people attending meetings and doing work?

675. Does the software engineering team have the right mix of skills?

676. Are there any HR Analytics projects similar to this one in existence?

677. Is security a central objective?

678. How would you assess the risk management process in the HR Analytics project?

679. Is a software HR Analytics project management tool available?

680. What is the probability of the risk occurring?

681. How solid is the HR Analytics projection of competitive reaction?

682. Would avoiding any of corresponding impact the HR Analytics projects chance of success?

683. Are the risk data timely and relevant?

684. Which functions, departments, and activities of your organization are going to be affected?

685. Does the software interface with new or unproven hardware or unproven vendor products?

686. What are the industrial relations prevailing in your organization?

687. Which role do you have in the HR Analytics project?

688. Have customers been involved fully in the definition of requirements?

2.34 Probability and Impact Matrix: HR Analytics

689. Which phase of the HR Analytics project do you take part in?

690. What should be the gestation period for the HR Analytics project with this technology?

691. During which risk management process is a determination to transfer a risk made?

692. How likely is the current plan to come in on schedule or on budget?

693. What are the uncertainties associated with the technology selected for the HR Analytics project?

694. What needs to be DONE?

695. Prioritized components/features?

696. Are HR Analytics project requirements stable?

697. Management -what contingency plans do you have if the risk becomes a reality?

698. Is the customer willing to establish rapid communication links with the developer?

699. Can it be changed quickly?

700. What are the preparations required for facing

difficulties?

701. What would be the best solution?

702. Is the HR Analytics project cutting across the entire organization?

703. What risks are necessary to achieve success?

704. What did not work so well?

705. How is the risk management process used in practice?

706. What are the probable external agencies to act as HR Analytics project manager?

2.35 Risk Data Sheet: HR Analytics

707. Whom do you serve (customers)?

708. Potential for recurrence?

709. During work activities could hazards exist?

710. Do effective diagnostic tests exist?

711. Who has a vested interest in how you perform as your organization (our stakeholders)?

712. What are the main threats to your existence?

713. How reliable is the data source?

714. Is the data sufficiently specified in terms of the type of failure being analyzed, and its frequency or probability?

715. Type of risk identified?

716. What can you do?

717. Will revised controls lead to tolerable risk levels?

718. What are you trying to achieve (Objectives)?

719. What will be the consequences if it happens?

720. What were the Causes that contributed?

721. What do people affected think about the need

for, and practicality of preventive measures?

722. What can happen?

723. What is the likelihood of it happening?

724. How can it happen?

725. What are the main opportunities available to you that you should grab while you can?

726. What are you here for (Mission)?

2.36 Procurement Management Plan: HR Analytics

727. Are quality inspections and review activities listed in the HR Analytics project schedule(s)?

728. Is there a procurement management plan in place?

729. Were HR Analytics project team members involved in the development of activity & task decomposition?

730. Have all involved HR Analytics project stakeholders and work groups committed to the HR Analytics project?

731. Is a pmo (HR Analytics project management office) in place which provides oversight to the HR Analytics project?

732. Are the people assigned to the HR Analytics project sufficiently qualified?

733. Is the schedule updated on a periodic basis?

734. What types of contracts will be used?

735. Are issues raised, assessed, actioned, and resolved in a timely and efficient manner?

736. How long will it take for the purchase cost to be the same as the lease cost?

737. Are the HR Analytics project team members located locally to the users/stakeholders?

738. What areas does the group agree are the biggest success on the HR Analytics project?

739. Are corrective actions and variances reported?

740. Are trade-offs between accepting the risk and mitigating the risk identified?

741. Is there a formal process for updating the HR Analytics project baseline?

742. Is the communication plan being followed?

2.37 Source Selection Criteria: HR Analytics

743. Do you ensure you evaluate what you asked for, not what you want to see or expect to see?

744. How should oral presentations be prepared for?

745. With the rapid changes in information technology, will media be readable in five or ten years?

746. What should be considered?

747. What should be considered when developing evaluation standards?

748. When is it appropriate to issue a Draft Request for Proposal (DRFP)?

749. Has all proposal data been loaded?

750. What risks were identified in the proposals?

751. What documentation is necessary regarding electronic communications?

752. Are types/quantities of material, facilities appropriate?

753. Do you have a plan to document consensus results including disposition of any disagreement by individual evaluators?

754. What does a sample rating scale look like?

755. Which contract type places the most risk on the seller?

756. Is the offeror pricing what is technically proposed?

757. Are there any specific considerations that precludes offers from being selected as the awardee?

758. What should communications be used to accomplish?

759. How should comments received in response to a RFP be handled?

760. Why promote competition?

761. Are considerations anticipated?

762. What aspects should the contracting officer brief the HR Analytics project on prior to evaluation of proposals?

2.38 Stakeholder Management Plan: HR Analytics

763. Do HR Analytics project managers participating in the HR Analytics project know the HR Analytics projects true status first hand?

764. Are communication systems currently in place appropriate?

765. What are reporting requirements?

766. Was trending evident between audits?

767. Are schedule deliverables actually delivered?

768. Who is responsible for arranging and managing the review(s)?

769. Have activity relationships and interdependencies within tasks been adequately identified?

770. Is the HR Analytics project sponsor clearly communicating the business case or rationale for why this HR Analytics project is needed?

771. How will the equipment be verified?

772. Which impacts could serve as impediments?

773. Are written status reports provided on a designated frequent basis?

774. Will the current technology alter during the life of the HR Analytics project?

775. What are the criteria for selecting suppliers of off the shelf products?

776. Where are the verification requirements to be documented (eg purchase order, agreement etc)?

777. What are the procedures and processes to be followed for purchases, including approval and authorisation requirements?

2.39 Change Management Plan: HR Analytics

778. What risks may occur upfront?

779. What do you expect the target audience to do, say, think or feel as a result of this communication?

780. Is there support for this application(s) and are the details available for distribution?

781. Has the training provider been established?

782. What prerequisite knowledge or training is required?

783. What is the most cynical response it can receive?

784. How prevalent is Resistance to Change?

785. What is the negative impact of communicating too soon or too late?

786. How might they respond to the message and if the response may be negative or open to misinterpretation, what else needs to be said?

787. How do you gain sponsors buy-in to the communication plan?

788. What is going to be done differently?

789. What type of materials/channels will be available

to leverage?

790. What did the people around you say about it?

791. Identify the current level of skills and knowledge and behaviours of the group that will be impacted on. What prerequisite knowledge do corresponding groups need?

792. Do you need a new organizational structure?

793. Where will the funds come from?

794. Do you need a new organization structure?

795. Who is the audience for change management activities?

796. Change invariability confront many relationships especially the already stated that require a set of behaviours What roles with in your organization are affected and how?

3.0 Executing Process Group: HR Analytics

797. How could stakeholders negatively impact your HR Analytics project?

798. When do you share the scorecard with managers?

799. How can software assist in HR Analytics project communications?

800. How could you control progress of your HR Analytics project?

801. How well did the chosen processes fit the needs of the HR Analytics project?

802. How can your organization use a weighted decision matrix to evaluate proposals as part of source selection?

803. What areas were overlooked on this HR Analytics project?

804. What were things that you need to improve?

805. What is the difference between conceptual, application, and evaluative questions?

806. Who are the HR Analytics project stakeholders?

807. Mitigate. what will you do to minimize the

impact should a risk event occur?

808. In what way has the program come up with innovative measures for problem-solving?

809. Contingency planning. if a risk event occurs, what will you do?

810. Does the case present a realistic scenario?

811. If a risk event occurs, what will you do?

812. How well did the team follow the chosen processes?

813. What are the HR Analytics project management deliverables of each process group?

814. It under budget or over budget?

815. Would you rate yourself as being risk-averse, risk-neutral, or risk-seeking?

3.1 Team Member Status Report: HR Analytics

816. Are the attitudes of staff regarding HR Analytics project work improving?

817. Will the staff do training or is that done by a third party?

818. Does your organization have the means (staff, money, contract, etc.) to produce or to acquire the product, good, or service?

819. What is to be done?

820. Why is it to be done?

821. How can you make it practical?

822. Do you have an Enterprise HR Analytics project Management Office (EPMO)?

823. How does this product, good, or service meet the needs of the HR Analytics project and your organization as a whole?

824. Are the products of your organizations HR Analytics projects meeting customers objectives?

825. The problem with Reward & Recognition Programs is that the truly deserving people all too often get left out. How can you make it practical?

826. How much risk is involved?

827. Does the product, good, or service already exist within your organization?

828. Does every department have to have a HR Analytics project Manager on staff?

829. Is there evidence that staff is taking a more professional approach toward management of your organizations HR Analytics projects?

830. When a teams productivity and success depend on collaboration and the efficient flow of information, what generally fails them?

831. What specific interest groups do you have in place?

832. Are your organizations HR Analytics projects more successful over time?

833. How it is to be done?

834. How will resource planning be done?

3.2 Change Request: HR Analytics

835. Who is responsible for the implementation and monitoring of all measures?

836. What mechanism is used to appraise others of changes that are made?

837. Can you answer what happened, who did it, when did it happen, and what else will be affected?

838. For which areas does this operating procedure apply?

839. Will all change requests and current status be logged?

840. How to get changes (code) out in a timely manner?

841. Describe how modifications, enhancements, defects and/or deficiencies shall be notified (e.g. Problem Reports, Change Requests etc) and managed. Detail warranty and/or maintenance periods?

842. Will this change conflict with other requirements changes (e.g., lead to conflicting operational scenarios)?

843. Can static requirements change attributes like the size of the change be used to predict reliability in execution?

844. Who is communicating the change?

845. Change request coordination ?

846. Will all change requests be unconditionally tracked through this process?

847. Who will perform the change?

848. What are the basic mechanics of the Change Advisory Board (CAB)?

849. What has an inspector to inspect and to check?

850. How can changes be graded?

851. What is a Change Request Form?

852. How are changes requested (forms, method of communication)?

853. Who is responsible to authorize changes?

854. How shall the implementation of changes be recorded?

3.3 Change Log: HR Analytics

855. Do the described changes impact on the integrity or security of the system?

856. When was the request approved?

857. Who initiated the change request?

858. Is the change request within HR Analytics project scope?

859. Is the requested change request a result of changes in other HR Analytics project(s)?

860. Is the submitted change a new change or a modification of a previously approved change?

861. When was the request submitted?

862. Where do changes come from?

863. Is the change backward compatible without limitations?

864. How does this relate to the standards developed for specific business processes?

865. Should a more thorough impact analysis be conducted?

866. How does this change affect the timeline of the schedule?

867. Is the change request open, closed or pending?

868. Is this a mandatory replacement?

869. Will the HR Analytics project fail if the change request is not executed?

870. How does this change affect scope?

3.4 Decision Log: HR Analytics

871. Is everything working as expected?

872. Meeting purpose; why does this team meet?

873. How do you define success?

874. How does an increasing emphasis on cost containment influence the strategies and tactics used?

875. What is the average size of your matters in an applicable measurement?

876. At what point in time does loss become unacceptable?

877. Is your opponent open to a non-traditional workflow, or will it likely challenge anything you do?

878. It becomes critical to track and periodically revisit both operational effectiveness; Are you noticing all that you need to, and are you interpreting what you see effectively?

879. What alternatives/risks were considered?

880. Who is the decisionmaker?

881. Does anything need to be adjusted?

882. What eDiscovery problem or issue did your organization set out to fix or make better?

883. How effective is maintaining the log at facilitating organizational learning?

884. What makes you different or better than others companies selling the same thing?

885. How consolidated and comprehensive a story can you tell by capturing currently available incident data in a central location and through a log of key decisions during an incident?

886. Behaviors; what are guidelines that the team has identified that will assist them with getting the most out of team meetings?

887. Linked to original objective?

888. Adversarial environment. is your opponent open to a non-traditional workflow, or will it likely challenge anything you do?

889. How do you know when you are achieving it?

890. Do strategies and tactics aimed at less than full control reduce the costs of management or simply shift the cost burden?

3.5 Quality Audit: HR Analytics

891. For each device to be reconditioned, are device specifications, such as appropriate engineering drawings, component specifications and software specifications, maintained?

892. How does your organization know that its system for managing intellectual property issues is appropriately effective, constructive and fair?

893. What data about organizational performance is routinely collected and reported?

894. Can your organization demonstrate exactly how and why results were achieved?

895. Are goals well supported with strategies, operational plans, manuals and training?

896. Is quality audit a prerequisite for program accreditation or program recognition?

897. How does your organization know that its relationships with the community at large are appropriately effective and constructive?

898. What are your supplier audits?

899. How does your organization know that its processes for managing severance are appropriately effective, constructive and fair?

900. How does your organization know that it is

maintaining a conducive staff climate?

901. How does your organization know that its system for recruiting the best staff possible are appropriately effective and constructive?

902. Does the suppliers quality system have a written procedure for corrective action when a defect occurs?

903. Has a written procedure been established to identify devices during all stages of receipt, reconditioning, distribution and installation so that mix-ups are prevented?

904. What is your organizations greatest strength?

905. Are storage areas and reconditioning operations designed to prevent mix-ups and assure orderly handling of both the distressed and reconditioned devices?

906. How does your organization know that its staff are presenting original work, and properly acknowledging the work of others?

907. How does your organization know that its public relations and marketing systems are appropriately effective and constructive?

908. How does the organization know that its system for maintaining and advancing the capabilities of its staff, particularly in relation to the Mission of the organization, is appropriately effective and constructive?

909. Have personnel cleanliness and health

requirements been established?

910. How does your organization know that the review processes are effective?

3.6 Team Directory: HR Analytics

911. Process decisions: do job conditions warrant additional actions to collect job information and document on-site activity?

912. How does the team resolve conflicts and ensure tasks are completed?

913. Who should receive information (all stakeholders)?

914. Who will report HR Analytics project status to all stakeholders?

915. Is construction on schedule?

916. Process decisions: do invoice amounts match accepted work in place?

917. Do purchase specifications and configurations match requirements?

918. Process decisions: which organizational elements and which individuals will be assigned management functions?

919. Have you decided when to celebrate the HR Analytics projects completion date?

920. Why is the work necessary?

921. Timing: when do the effects of communication take place?

922. Who will be the stakeholders on your next HR Analytics project?

923. Who are the Team Members?

924. Process decisions: are contractors adequately prosecuting the work?

925. Decisions: what could be done better to improve the quality of the constructed product?

926. How and in what format should information be presented?

927. When will you produce deliverables?

928. How do unidentified risks impact the outcome of the HR Analytics project?

929. What are you going to deliver or accomplish?

3.7 Team Operating Agreement: HR Analytics

930. What are the current caseload numbers in the unit?

931. Are there differences in access to communication and collaboration technology based on team member location?

932. Confidentiality: how will confidential information be handled?

933. How will you resolve conflict efficiently and respectfully?

934. Seconds for members to respond?

935. The method to be used in the decision making process; Will it be consensus, majority rule, or the supervisor having the final say?

936. Must your team members rely on the expertise of other members to complete tasks?

937. Are there more than two functional areas represented by your team?

938. Do you listen for voice tone and word choice to understand the meaning behind words?

939. What is culture?

940. Did you draft the meeting agenda?

941. Do you call or email participants to ensure understanding, follow-through and commitment to the meeting outcomes?

942. Do you determine the meeting length and time of day?

943. Is compensation based on team and individual performance?

944. Do you ask participants to close laptops and place mobile devices on silent on the table while the meeting is in progress?

945. Must your members collaborate successfully to complete HR Analytics projects?

946. Do you solicit member feedback about meetings and what would make them better?

947. What is your unique contribution to your organization?

948. Are leadership responsibilities shared among team members (versus a single leader)?

3.8 Team Performance Assessment: HR Analytics

949. Where to from here?

950. To what degree will the team ensure that all members equitably share the work essential to the success of the team?

951. Does more radicalness mean more perceived benefits?

952. To what degree do members articulate the goals beyond the team membership?

953. What structural changes have you made or are you preparing to make?

954. To what degree do team members feel that the purpose of the team is important, if not exciting?

955. To what degree do team members agree with the goals, relative importance, and the ways in which achievement will be measured?

956. How do you recognize and praise members for contributions?

957. Which situations call for a more extreme type of adaptiveness in which team members actually re-define roles?

958. To what degree can team members frequently

and easily communicate with one another?

959. To what degree is the team cognizant of small wins to be celebrated along the way?

960. Social categorization and intergroup behaviour: Does minimal intergroup discrimination make social identity more positive?

961. To what degree will the approach capitalize on and enhance the skills of all team members in a manner that takes into consideration other demands on members of the team?

962. When a reviewer complains about method variance, what is the essence of the complaint?

963. To what degree can the team measure progress against specific goals?

964. To what degree is there a sense that only the team can succeed?

965. To what degree do team members articulate the teams work approach?

966. How hard do you try to make a good selection?

967. To what degree does the teams purpose constitute a broader, deeper aspiration than just accomplishing short-term goals?

3.9 Team Member Performance Assessment: HR Analytics

968. What are the basic principles and objectives of performance measurement and assessment?

969. Are the draft goals SMART ?

970. What evidence supports your decision-making?

971. Is there reluctance to join a team?

972. How accurately is your plan implemented?

973. How often should assessments be conducted?

974. What, if any, steps are available for employees who feel they have been unfairly or inaccurately rated?

975. To what degree are the goals realistic?

976. What are they responsible for?

977. What is the Business Management Oversight Process?

978. Does adaptive training work?

979. What evaluation results do you have?

980. How is your organizations Strategic Management System tied to performance measurement?

981. How do you make use of research?

982. In what areas would you like to concentrate your knowledge and resources?

983. What specific plans do you have for developing effective cross-platform assessments in a blended learning environment?

3.10 Issue Log: HR Analytics

984. Are there potential barriers between the team and the stakeholder?

985. Are the HR Analytics project issues uniquely identified, including to which product they refer?

986. Why multiple evaluators?

987. How often do you engage with stakeholders?

988. Why do you manage human resources?

989. Who have you worked with in past, similar initiatives?

990. Can an impact cause deviation beyond team, stage or HR Analytics project tolerances?

991. What would have to change?

992. Why not more evaluators?

993. In classifying stakeholders, which approach to do so are you using?

994. How much time does it take to do it?

995. Who reported the issue?

996. Is access to the Issue Log controlled?

997. Are they needed?

998. How do you manage communications?

4.0 Monitoring and Controlling Process Group: HR Analytics

999. Change, where should you look for problems?

1000. Have operating capacities been created and/or reinforced in partners?

1001. User: who wants the information and what are they interested in?

1002. What areas were overlooked on this HR Analytics project?

1003. Were decisions made in a timely manner?

1004. What were things that you did very well and want to do the same again on the next HR Analytics project?

1005. What kinds of things in particular are you looking for data on?

1006. Did the HR Analytics project team have enough people to execute the HR Analytics project plan?

1007. Is there undesirable impact on staff or resources?

1008. Use: how will they use the information?

1009. How is agile HR Analytics project management done?

1010. Based on your HR Analytics project communication management plan, what worked well?

1011. What business situation is being addressed?

1012. Propriety: who needs to be involved in the evaluation to be ethical?

1013. Who needs to be involved in the planning?

1014. What factors are contributing to progress or delay in the achievement of products and results?

1015. How is agile portfolio management done?

1016. What resources (both financial and non-financial) are available/needed?

4.1 Project Performance Report: HR Analytics

1017. What is the PRS?

1018. To what degree do team members understand one anothers roles and skills?

1019. To what degree does the teams work approach provide opportunity for members to engage in results-based evaluation?

1020. To what degree can the team ensure that all members are individually and jointly accountable for the teams purpose, goals, approach, and work-products?

1021. To what degree does the information network provide individuals with the information they require?

1022. To what degree are the skill areas critical to team performance present?

1023. To what degree can the cognitive capacity of individuals accommodate the flow of information?

1024. To what degree are the tasks requirements reflected in the flow and storage of information?

1025. To what degree does the informal organization make use of individual resources and meet individual needs?

1026. To what degree will the team adopt a concrete, clearly understood, and agreed-upon approach that will result in achievement of the teams goals?

1027. How will procurement be coordinated with other HR Analytics project aspects, such as scheduling and performance reporting?

1028. To what degree does the funding match the requirement?

1029. To what degree do the structures of the formal organization motivate taskrelevant behavior and facilitate task completion?

1030. To what degree does the teams approach to its work allow for modification and improvement over time?

1031. To what degree does the team possess adequate membership to achieve its ends?

4.2 Variance Analysis: HR Analytics

1032. Are there changes in the overhead pool and/or organization structures?

1033. What are the direct labor dollars and/or hours?

1034. Why are standard cost systems used?

1035. Are all budgets assigned to control accounts?

1036. What business event causes fluctuations?

1037. Does the scheduling system identify in a timely manner the status of work?

1038. What causes selling price variance?

1039. How does your organization allocate the cost of shared expenses and services?

1040. What should management do?

1041. How are variances affected by multiple material and labor categories?

1042. What is the expected future profitability of each customer?

1043. How does the use of a single conversion element (rather than the traditional labor and overhead elements) affect standard costing?

1044. Contemplated overhead expenditure for each

period based on the best information currently is available?

1045. Are the overhead pools formally and adequately identified?

1046. What types of services and expense are shared between business segments?

1047. When, during the last four quarters, did a primary business event occur causing a fluctuation?

1048. Do the rates and prices remain constant throughout the year?

1049. Are meaningful indicators identified for use in measuring the status of cost and schedule performance?

1050. Did a new competitor enter the market?

4.3 Earned Value Status: HR Analytics

1051. Where is evidence-based earned value in your organization reported?

1052. Verification is a process of ensuring that the developed system satisfies the stakeholders agreements and specifications; Are you building the product right? What do you verify?

1053. How much is it going to cost by the finish?

1054. What is the unit of forecast value?

1055. Validation is a process of ensuring that the developed system will actually achieve the stakeholders desired outcomes; Are you building the right product? What do you validate?

1056. Are you hitting your HR Analytics projects targets?

1057. If earned value management (EVM) is so good in determining the true status of a HR Analytics project and HR Analytics project its completion, why is it that hardly any one uses it in information systems related HR Analytics projects?

1058. When is it going to finish?

1059. Where are your problem areas?

1060. Earned value can be used in almost any HR Analytics project situation and in almost any HR

Analytics project environment. it may be used on large HR Analytics projects, medium sized HR Analytics projects, tiny HR Analytics projects (in cut-down form), complex and simple HR Analytics projects and in any market sector. some people, of course, know all about earned value, they have used it for years - but perhaps not as effectively as they could have?

1061. How does this compare with other HR Analytics projects?

4.4 Risk Audit: HR Analytics

1062. Who is responsible for what?

1063. How risk averse are you?

1064. Are policies communicated to all affected?

1065. Are the best people available?

1066. Do you have an emergency plan?

1067. Do you promote education and training opportunities?

1068. Should additional substantive testing be conducted because of the risk audit results?

1069. Are testing tools available and suitable?

1070. Does your organization have or has considered the need for insurance covers: public liability, professional indemnity and directors and officers liability?

1071. Does the adoption of a business risk audit approach change internal control documentation and testing practices?

1072. Do you have a mechanism for managing change?

1073. What programmatic and Fiscal information is being collected and analyzed?

1074. Is your organization an exempt employer for payroll tax purposes?

1075. Is the auditor able to evaluate contradictory evidence in an unbiased manner?

1076. When your organization is entering into a major contract, does it seek legal advice?

1077. Are procedures developed to respond to foreseeable emergencies and communicated to all involved?

1078. What effect would a better risk management program have had?

1079. Does your organization have a process for meeting its ongoing taxation obligations?

1080. Level of preparation and skill?

1081. Are all managers or operators of the facility or equipment competent or qualified?

4.5 Contractor Status Report: HR Analytics

1082. What are the minimum and optimal bandwidth requirements for the proposed solution?

1083. How is risk transferred?

1084. Who can list a HR Analytics project as organization experience, your organization or a previous employee of your organization?

1085. What process manages the contracts?

1086. What was the final actual cost?

1087. What was the budget or estimated cost for your organizations services?

1088. How long have you been using the services?

1089. What was the actual budget or estimated cost for your organizations services?

1090. What was the overall budget or estimated cost?

1091. Are there contractual transfer concerns?

1092. What is the average response time for answering a support call?

1093. Describe how often regular updates are made to the proposed solution. Are corresponding regular updates included in the standard maintenance plan?

1094. If applicable; describe your standard schedule for new software version releases. Are new software version releases included in the standard maintenance plan?

1095. How does the proposed individual meet each requirement?

4.6 Formal Acceptance: HR Analytics

1096. What are the requirements against which to test, Who will execute?

1097. What lessons were learned about your HR Analytics project management methodology?

1098. Was business value realized?

1099. Was the sponsor/customer satisfied?

1100. What is the Acceptance Management Process?

1101. How well did the team follow the methodology?

1102. Was the HR Analytics project managed well?

1103. What function(s) does it fill or meet?

1104. Was the HR Analytics project goal achieved?

1105. Was the client satisfied with the HR Analytics project results?

1106. Do you buy pre-configured systems or build your own configuration?

1107. Do you buy-in installation services?

1108. Did the HR Analytics project achieve its MOV?

1109. Who supplies data?

1110. Have all comments been addressed?

1111. Does it do what client said it would?

1112. Does it do what HR Analytics project team said it would?

1113. General estimate of the costs and times to complete the HR Analytics project?

1114. What was done right?

1115. Who would use it?

5.0 Closing Process Group: HR Analytics

1116. How dependent is the HR Analytics project on other HR Analytics projects or work efforts?

1117. What was learned?

1118. What were things that you did well, and could improve, and how?

1119. Is this an updated HR Analytics project Proposal Document?

1120. How well defined and documented were the HR Analytics project management processes you chose to use?

1121. Were escalated issues resolved promptly?

1122. What can you do better next time, and what specific actions can you take to improve?

1123. What areas were overlooked on this HR Analytics project?

1124. Will the HR Analytics project deliverable(s) replace a current asset or group of assets?

1125. Is there a clear cause and effect between the activity and the lesson learned?

1126. Did you do things well?

1127. What were things that you did very well and want to do the same again on the next HR Analytics project?

1128. How well did the chosen processes fit the needs of the HR Analytics project?

1129. What is the HR Analytics project Management Process?

1130. Were cost budgets met?

5.1 Procurement Audit: HR Analytics

1131. Do you learn from benchmarking your own practices with international standards?

1132. Is it calculated whether aggregated procurement can be more cost-efficient?

1133. Are the internal control systems operational?

1134. Are services/tasks combined in such a way that the market is used where relevant?

1135. Does the procurement function/unit have the ability to apply electronic procurement?

1136. Can changes be made to automatic disbursement programs without proper approval of management?

1137. Is your organization aware and informed about international procurement standards and good practice?

1138. Is the issuance of purchase orders scheduled so that orders are not issued daily?

1139. Is there a need for the procurement HR Analytics project at all?

1140. Access to data, including standing data, and the identification of restriction levels and authorised personnel was in place?

1141. Is there time waste during tendering?

1142. Does procurement staff have recognized professional procurement qualifications or sufficient training?

1143. Are reports based on sound data available to the already stated responsible for monitoring the performance of contracts?

1144. Are information gathered to produce knowledge about procured goods and services, prices paid and supplier performance?

1145. Are there complementary rules to be used and are they applied?

1146. Proper and complete records of transactions and events are maintained?

1147. Are open purchase orders with a fixed monetary limitation used for local purchases of small dollar value?

1148. Has the award included no items different from the already stated contained in bid specifications?

1149. If information was withheld, was there reasonable justification for this decision?

1150. Are the rules for automatic payment in computer programs approved by management prior to implementation?

5.2 Contract Close-Out: HR Analytics

1151. Was the contract type appropriate?

1152. Change in attitude or behavior?

1153. Change in knowledge?

1154. How does it work?

1155. Have all contracts been completed?

1156. Why Outsource?

1157. Are the signers the authorized officials?

1158. Has each contract been audited to verify acceptance and delivery?

1159. Have all contract records been included in the HR Analytics project archives?

1160. What is capture management?

1161. Parties: Authorized?

1162. Was the contract complete without requiring numerous changes and revisions?

1163. How is the contracting office notified of the automatic contract close-out?

1164. Have all contracts been closed?

1165. What happens to the recipient of services?

1166. How/when used ?

1167. Was the contract sufficiently clear so as not to result in numerous disputes and misunderstandings?

1168. Have all acceptance criteria been met prior to final payment to contractors?

1169. Parties: who is involved?

1170. Change in circumstances?

5.3 Project or Phase Close-Out: HR Analytics

1171. Does the lesson educate others to improve performance?

1172. Planned completion date?

1173. What was the preferred delivery mechanism?

1174. What are the informational communication needs for each stakeholder?

1175. Were the outcomes different from the already stated planned?

1176. Is the lesson based on actual HR Analytics project experience rather than on independent research?

1177. What are the marketing communication needs for each stakeholder?

1178. What process was planned for managing issues/risks?

1179. Does the lesson describe a function that would be done differently the next time?

1180. What were the goals and objectives of the communications strategy for the HR Analytics project?

1181. What hierarchical authority does the

stakeholder have in your organization?

1182. What is this stakeholder expecting?

1183. What was expected from each stakeholder?

1184. What benefits or impacts does the stakeholder group expect to obtain as a result of the HR Analytics project?

1185. Who controlled key decisions that were made?

1186. Who is responsible for award close-out?

1187. Planned remaining costs?

1188. What are the mandatory communication needs for each stakeholder?

1189. Did the delivered product meet the specified requirements and goals of the HR Analytics project?

5.4 Lessons Learned: HR Analytics

1190. What could have been improved?

1191. How well defined were the acceptance criteria for HR Analytics project deliverables?

1192. Were the HR Analytics project objectives met (if not, briefly account for what wasnt met)?

1193. How much of your time was spent on other than this HR Analytics project?

1194. What is the growth stage of the organization?

1195. Were the right people available when required?

1196. What is in the future?

1197. What report generation capability is needed?

1198. Which estimation issues did you personally have and what was the impact?

1199. How well were your expectations met regarding the extent of your involvement in the HR Analytics project (effort, time commitments, etc.)?

1200. What is your working hypothesis, if you have one?

1201. Is the lesson based on actual HR Analytics project experience rather than on independent research?

1202. What HR Analytics project circumstances were not anticipated?

1203. How well did the scope of the HR Analytics project match what was defined in the HR Analytics project Proposal?

1204. How mature are the observations?

1205. What is your strategy for data collection?

1206. How effectively were issues resolved before escalation was necessary?

1207. What is your overall assessment of the outcome of this HR Analytics project?

1208. Did the team work well together?

1209. What is the expected lifespan of the deliverable?

Index

Made in United States
Orlando, FL
04 November 2026

90205883R00182